S. M Saxby

The Projection and Calculation of the Sphere

S. M Saxby

The Projection and Calculation of the Sphere

ISBN/EAN: 9783337396923

Printed in Europe, USA, Canada, Australia, Japan

Cover: Foto ©berggeist007 / pixelio.de

More available books at **www.hansebooks.com**

THE

PROJECTION AND CALCULATION

OF

THE SPHERE

FOR YOUNG SEA OFFICERS

BEING

A COMPLETE INITIATION INTO NAUTICAL ASTRONOMY

BY S. M. SAXBY, R.N.

PRINCIPAL INSTRUCTOR OF NAVAL ENGINEERS, HER MAJESTY'S STEAM RESERVE
LATE OF CAIUS COLLEGE, CAMBRIDGE

LONDON
LONGMAN, GREEN, LONGMAN, AND ROBERTS
1861 .

PREFACE.

SPHERIC TRIGONOMETRY is so little understood by those who are not professionally mathematicians, while the subject is so important to a maritime nation, that the question whether there be not defect in the present mode of educating nautical men demands earnest attention.

For more than a century the method of teaching what is called the "Art of Navigation" has been gradually changing, and the knowledge of *principles* has in England been consequently decreasing, while other nations, possessed of a navy and commercial marine vastly inferior to those of England, have maintained a standard which actually originated in the labours and experience of our own early nautical teachers.

It is remarkable that, while in the year 1700 no one who was ignorant of Spherics, and the principles of Nautical Astronomy, was considered a competent navigator, in the year 1860, the "epitomes" of navigation used both in the Royal Navy and the Merchant Service of this great country, not only make no pretension to the teaching of Spherics, either by calculation or construction, but one of

A 4

them (which on all other naval subjects is elaborate and accurate) even repudiates the propriety of mixing theory with practice, while certain other works insist on theory as the only medium by which to obtain a correct practical knowledge of the subject.

The true system of teaching lies between these extremes; and this little work is an attempt to facilitate the study by sufficiently explaining general principles while adapting them to immediate practice.

A few recent works could be referred to which have already, by an advance in the right direction, whetted the mental appetite for really digestible and nourishing sustenance; but the young naval officer and the young aspiring merchant commander of the present generation have absolutely (beyond these) no available means of improving their acquaintance with what they feel to be indisputably essential to their self-respect as navigators, inasmuch as works in publication either disregard theory altogether as unimportant to a beginner, or bewilder the student with a phalanx of formulæ, which in their very aspect so commonly suggest a series of difficulties insurmountable, that relapse into indifference is the natural consequence.

And yet naval instructors are expected to impart a knowledge of Nautical Astronomy adequate to the needs of a sailor. Such is undoubtedly done to the very limit of their powers, but few besides nautical teachers can appreciate the labour of the work of elucidating a subject like Nautical Astronomy, in the absence of well-digested plans of instruction in a printed form, in which everything would be explained *so clearly as to encourage the student, and lead him step by step up the ladder of knowledge, when the ascent is to be attempted away from professional assistance.*

The subject of Spherics is one of such profound research and extent, that it seems to have been expected that a certain conciseness of language and solemnity of tone would best become the author who would enter publicly upon its teaching. But *familiar* description, divested of "hard words," has been so successfully attempted by an excellent mathematician of our own ·times while treating of sister branches of science, that the example of such men as the Rev. Harvey Goodwin (whose illustrations of general mathematics have so much benefited the Cambridge student) may safely be imitated; certainly by one, it is presumed, whose experience as a teacher of adults extends over a period of some three and thirty years.

There may be about 100,000 navigators in Great Britain, very likely there are considerably more, inasmuch as in the merchant service alone there are about 45,000 vessels afloat. It is scarcely, however, to be expected that the "wary mariner," trained as he has been to vigilance, and ever alive as he consequently is to necessity for precautions in his sea voyage, will not apply the same to his voyage of life. *It is contrary to his professional habits* to venture heedlessly among the rocks and shoals which may beset a coast to him unknown; and in like manner such are his precautions with regard to any *science* to him unknown; but give him even a tolerably clear *outline* "chart," and general "sailing directions," or even furnish him with a "hand lead" wherewith he may *feel the bottom if he cannot see it*, and there are in his occupations days and hours of thoughtful leisure in which he may be tempted to vary

his "cruise," and at least examine for himself the creeks
and lagoons of science. At present these are obscured
from his vision by characteristic prejudices and distrust.
To remove these is then the object of this little book.
Whatever may be the beauties of a science so captivating
as Nautical Astronomy, the very approaches have been
made toilsome from the *decay of the landmarks* which
once guided the young seaman on his path. And when a
straggling, casual, and mere wayfaring adventurer has
accidentally gained a "peep" within the barriers, what
has he seen but a frowning array of sines, secants, tangents,
&c., twisted into every imaginable equational and fractional
form, and distorted into a thousand labyrinthine channels,
leading into deeps to him unfathomable. It is in vain to
say such is a befitting initiation for a young aspirant in
Nautical Science. It appals him; it drives him back to
his ordinary level, discourages from other attempts to advance himself in the scale of proficiency, and throws him
upon the dangerous and debilitating "consolation" that
among his own class he can still "pass muster." Many
indefatigable navigators *do* however brave the difficulties
which present themselves, and attain a very fair footing :
the object of this book is to vastly increase their number.

It may be asked, could one in a thousand even illustrate
geometrically or by "projection" the simple question of
latitude from a meridian altitude, or answer it by the use
of scale and dividers ? The writer would be very sorry to
be compelled to publish, even from his individual experience, details of positive danger which this state of
things has entailed upon the commercial world; but one
thing is certain, viz. the importance of the interests of

maritime commerce is involved in the *safe transit* of its vast treasures; such interests are paramount, and demand our best efforts in their support, — a sufficient apology for this attempt.

It is not enough that naval officers have peculiar advantages in nautical training, or that certain inducements are proffered. by the Board of Trade to men capable of acquiring, or willing to contend for, " extra certificates " in the merchant service; for not even with the latter is a knowledge of Spherics expected beyond its one simple application to great circle-sailing, and *this without any requirements whatever as to calculation.* Hence we may safely affirm that the subject of teaching Spherics seems to demand complete revision.

Nor is it sufficient that the nautical tutors of the present day perform their duties in a manner which has already loosened, as it were, the stability of an erroneous system of teaching, and which would in time restore that system to its previous sturdy and useful basis. But it is desirable to shorten such time of changeful probation, and an endeavour, in order to be successful, must be bold and radical. Navigation in these days is not as formerly required for the slow hulls of the beginning of the last century, which averaged from four to six knots an hour, but for steamers, swiftly flying against wind and tide at the rate of — (it is scarcely prudent to say *what!*). A more *ready system* of navigation is therefore called for; and if years since it were important to the venerable structures to which allusion is made, that *they* should be navigated in the best possible method, by so much the more is it necessary that in this improved age of "clippers," all the efficacy of a *sound knowledge of principles* should be called into operation.

If Spherics *as a basis* of all oversea navigation was considered and found to be indispensable to the safe conduct of ships in 1680, by so much the more must a knowledge of it be essential to shipmasters in 1860.

In tracing the cause of the deterioration which the system of nautical training has undergone during the last century and a half, we shall find it shown in the preface to the still admirable and elaborate " Practice of Navigation," written by the lamented and unrequited Lieut. Raper, R.N., wherein he states that the theory and the practice are kept "*purposely distinct;*" and he adds " it is the custom generally to teach the theory first; the impression forced on me is on the contrary, that the practice is itself the best foundation for sound and rapid advancement in the theory. For he who has acquired the practice knows the nature and extent of the subject, and in proceeding to the theory he has a distinct perception of the object to be attained."

The author of this little work concurs fully in the views of Raper, wherein he advocates the all importance of practical knowledge to the student in theory; but the exclusion of theory is merely the lesser evil, and has led to the " rule of thumb " system, to the final exclusion of elucidation of principles from our best existing works on navigation on the one hand, or has engrossed it too exclusively on the other, in works intended for young sea officers; its inevitable consequence is therefore the state of things which it is the object of the writer of this to combat.

The mind of a seafaring man is so much occupied with responsibilities of duty, or as regards passengers, the crew, &c., that he has in general too little leisure for deep studies, and when on shore his necessary and reasonable repose admits of little interruption; therefore to benefit the whole class a work is wanted which, in familiar language, and in a proper blending of theory with practice can lead to a solid and respectable acquirement. The more easy the steps of knowledge can be rendered to so peculiarly situated a class as sea officers, the faster progress will he make who aims at a higher intellectual level; and, indeed, a very *gently inclined plane of science is more necessary for the navigator than for any other class of individuals.*

Such, then, this book is intended to be; such, indeed, as a young seaman may take up with a firm conviction that all essentials are fully explained; and with an interest in believing that if it be read even with the attention usually paid to a work of fiction, *as a mere pastime,* sound and valuable information will infallibly take root in his mind.

The oft-quoted assertion attributed to Euclid in his reply to one of the Ptolemies, viz. "that there is no royal road to learning," must no longer obtain among us. Not every seaman aspires to become an Archimedes, and if he did, an Alexandrian school might not be the only one and the best in which to form him.

What Sir John Herschel has done for astronomy, Faraday for chemistry, Sedgwick for geology, Arnold for classics, Hutton and Colenso for arithmetic and algebra, and Goodwin, Snowball, &c., for the study of mechanics,

may surely be done for the navigator.   Hence the object of
this book.

Ill understood, isolated and forbidding formulæ thrust
imperiously upon an already burdened memory are a load
and an oppression, while pleasant, easy illustration of
principles is an enticement, and wins upon the mind until
its powers are secured by the silken bonds of a willing
captivity.

The British Government are seconding the community
in their struggles for a better *middle class* education.
Shall then those whose perilous avocations demand espe-
cial intelligence, and to whose hands and *heads* we trust
our lives and properties, and above all our national honour
and the defence of our hearths, shall *these alone* be ex-
cluded from the social arena of progress, and as mere
spectators see the palm of proficiency in the grasp of a
foreigner?

                                              S. M. S.

CONTENTS.

TRIGONOMETRY

AND

NAUTICAL ASTRONOMY.

WHILE plane trigonometry in its application to the practice of navigation, is so well set forth in the usual epitomes and other nautical works, that little need be said herein upon the subject, beyond a few remarks on certain of its fundamental principles, no one can properly comprehend even the very elementary principles of nautical astronomy, without a better insight into spherical trigonometry than is to be obtained from any work at present accessible to the sea officer.

PROJECTION.

The study of spherics is too often undertaken in ignorance of a method of constructing a spheric angle by scale. Such methods (for there are several) are called "projections," a general term in spherics, signifying the transferring of spaces from rounded surfaces to flat or "plane" ones, in the forms in which the eye would trace them, according to their assumed relative position.

Distortion must of necessity accompany the projection

B

of a spheric surface upon a plane, and all methods used in accomplishing this are not equally good.

1. For example, the gnomonic projection presents peculiar advantages in dialling, and also in great circle or tangent sailing; while the orthographic projection is used by the astronomer in the delineation of eclipses, the transits of heavenly bodies, &c. Mercator's is exceedingly useful in the construction of sea charts; while the stereographic is more applicable than all others to the purposes of nautical astronomy, in consequence of all its parts being either arcs of circles or straight lines. The globular is used by map engravers, the scenographic for perspective, and need not occupy the attention of the nautical astronomer.

2. Our consideration, then, will be restricted to the orthographic, Mercator's and stereographic projections; merely first noting that in the gnomonic projection, the eye is supposed to be at the centre of the sphere, viewing the meridians as straight lines; and in this projection the shortest distance on the globe between two places is represented by the shortest distance between the two corresponding points on the flat surface.

ORTHOGRAPHIC.

3. In the orthographic projection, we, for the occasion, must suppose the eye to be placed at an immense distance,

Fig. 1.

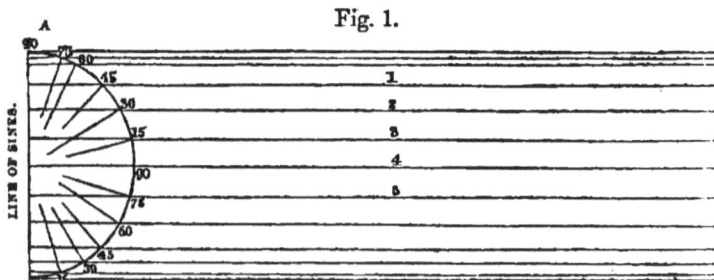

and as viewing a sphere as if it were a mere disc, such as
the sun or moon appears to be, the visual rays, as 1, 2, 3,
&c., in Fig. 1, being supposed to be parallel to each other;
in such case parallels of latitude in a right sphere would
appear (as *c, d, x,* in Fig. 2) to be straight lines, and separated in the proportion of the divisions on what we
call the line of sines upon the scale; while the meridians
lying near the diameter would appear to be at nearly
their actual distances asunder, but would be crowded as
they lie nearer to the periphery as at *a* (Fig. 1).*

4. Supposing, further, that a globe of immense size had
the usual lines of the sphere marked upon it, it would, as

Fig. 2.

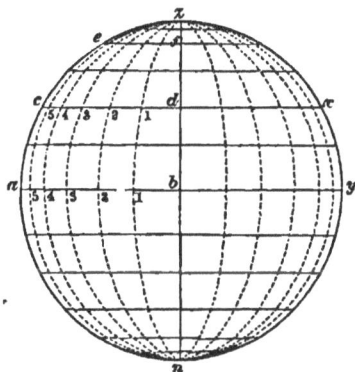

seen by the eye at (if possible) an extreme distance, have
the appearance of Fig. 2, in which all the meridians are
ellipses and *not* arcs of circles; this would readily be understood by holding a small toy globe at arm's length,
with the polar diameter in a vertical direction.

5. There is, perhaps, at the present time, no work in pub-

* Many illustrations might be given from one crowded diagram; but as
plainness is so desirable in a work like this, it is better to give separate
diagrams with the text therewith connected.

lication which would much assist the ordinary student in projecting the orthographic sphere. The methods given, and the explanations accompanying them, are beyond the comprehension of a beginner ; and are, moreover, so sufficiently troublesome to the practised draughtsman as to cause him in most cases merely to mark off the orthographic distances of the meridians on *the equator only,* and draw them as arcs of circles. This may even be seen in the diagrams illustrating the works of our greatest philosophers. Hence orthographic plates will seldom bear the inspection of the mathematician. It is not therefore extraordinary that the orthographic projection has been so little used for nautical purposes, until the writer suggested its introduction (as it will be further explained), to relieve the stereographic projection of certain disabilities in its application to the purposes of navigation.

6. The following directions for constructing an orthographic hemisphere, in what is called a " right sphere " (having the poles in the *circumference* as distinguished from a " parallel sphere " in which the poles are at the *centre,* and from an " oblique sphere " in which the poles are *neither* at the circumference nor at the centre) are given in detail, the method not having been yet published, and being the one long used by the writer. Its utility may at once be inferred from the circumstance of its requiring neither tangents nor secants, but simply the line of sines and scale of chords. It should be borne in mind that in Fig. 2 the parallels all appear as straight lines, like the equator; and if the line, *a b,* in its divisions forms the line of sines as seen at the diameter of Fig. 1, any corresponding portions of the other parallels must do the same, because they are supposed to be viewed from the same point; hence, *c d,* and *e f,* Fig. 2, &c., are each crossed by meridians, at distances proportional to the natural sines of their respective latitudes. In other words, $d\,2$ would

be a semi-minor axis of the meridian $z\,2\,n$, or the cosine
of its meridian. Every nautical astronomer, therefore,
will do well to have a scale prepared for general use,
similar to the one at Fig. 3. It is formed by making the
lines B C and B A meet at any angle (the nearer 90° the
more convenient), and laying off on B A the distances
B 15, B 30, &c., to B 90, from any scale of sines; then join
C 15, C 30, &c., and *parallel* to B A draw any number of
lines through the figure, and the scale is complete.

Fig. 2 may be drawn from this scale in the following
manner :—

Take A B from the scale (Fig. 3), as a radius, and describe a circle, drawing two diameters right-angled at the
centre. Take a straight edge of paper and lay it at B A,
Fig. 3; transfer on to it the distances, B 15, B 30, &c., up

Fig. 3.

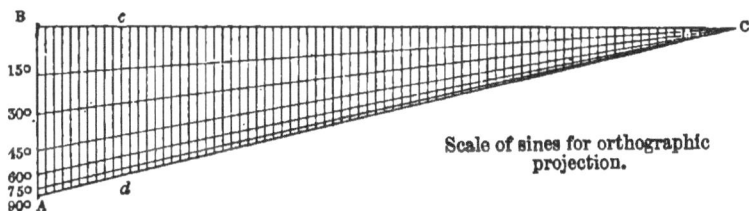

Scale of sines for orthographic
projection.

to 90 ; lay these off on Fig. 2, as $b\,1$, $b\,2$, &c., up to $b\,a$,
and also from b towards z, from b to y, and from b to n.
Through $b\,z$ and $b\,n$ draw parallels as in Fig. 2; then
take the length of any *half* parallels, as $c\,d$, on a straight
edge of paper, find its distance on the scale (Fig. 3), as
at $c\,d$, and having copied the divisions (as was done at B A
in dividing $a\,b$, Fig. 2), lay off these on Fig. 2, as $d\,1$,
$d\,2$, $d\,3$, &c. : proceed thus with each half parallel, and
points will be accurately obtained through which the whole
of the desired meridians may be constructed.

D 3

7. If it be required to project the sphere on the plane of the equator, or the "parallel sphere," proceed as follows. From the scale of sines (Fig. 3) take B A as radius, and with it draw a circle *a b c d*, quartering it, as before, at right angles, as in Fig. 4. If it be desirable to draw meridians through every 10°, or every 15° (suppose the latter) take the division of the line of sines, Fig. 3 (as before), apply them from *e* towards *b*, in Fig. 4; and with the centre *e*, and distances *e* 1, *e* 2, &c., describe circles; then lay off from *a* to *b*, *b* to *c*, *c* to *d*, and *d* to *a*, every 15th degree from a scale of chords corresponding

Fig. 4.

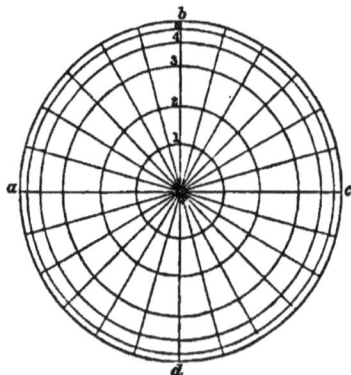

to the radius (or by dividing either quadrant into six equal parts); join these several points with the centre, and these form the meridians, the poles of the primitive circle being at the centre.

8. This mode of construction is particularly useful in projecting places in the polar regions, or places of heavenly bodies when near the poles of the ecliptic or equator; in such cases the least distortions being at and near the centre of the figure; but is most especially valuable in general questions of nautical astronomy, where the data lie near the prime vertical, such as altitudes, &c., at hours which are near 6 A.M. or 6 P.M., while for other periods of the day or night nearer to noon or midnight, the stereographic projection has its advantages.

MERCATOR'S.

9. Mercator's projection is too well known by nautical men to require much mention here, but its properties may be thus very briefly stated. On a globe it will be noticed that the actual measured length of degrees of longitude diminishes as they recede from the equator towards either pole.

In about the year 1590, a Mr. Wright, of Caius College, Cambridge, conceived the notion that he could on paper conveniently compensate this contraction of the degrees of longitude (in placing the round upon a flat surface), by *expanding* each degree of latitude in proportion to its corresponding degree of longitude in that latitude; by this means he projected the meridians and parallels as straight lines; hence, in this all rhumbs (or compass bearings) cross the meridians at equal angles, and a ship's course is laid down as a straight line. But this projection greatly distorts the outline and figures of places lying far from the equator. The miles in a degree of latitude are to the miles in a degree of longitude as radius is to cosine of latitude in which the degree of longitude is situated. For example :—Required the length in measured nautical miles of a degree of longitude in latitude 50°.

In Fig. 5, making hypothenuse radius, we have (by plane trigonometry)

rad $AC : 60 :: \cos 50° : AB = 38'\cdot57$ miles of long;

or in Fig. 6, making base radius

    sec 50° : 60 : : rad A B : 38'·57 miles of long.

Fig. 5.   Fig. 6

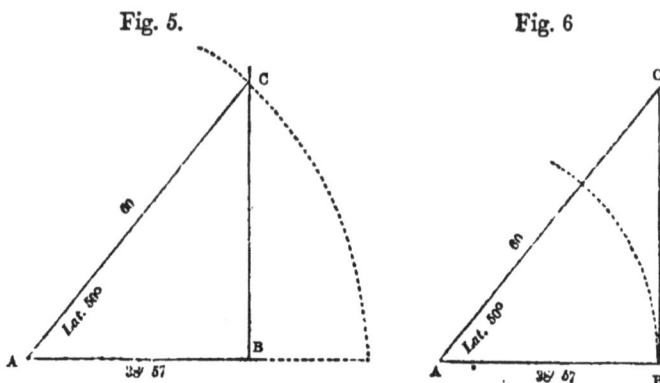

STEREOGRAPHIC.

10. The stereographic projection (which every navigator will feel the greatest pleasure and advantage in
comprehending if he read the following with ordinary
attention), enters largely into the daily work and interest
of every one who has command of a ship.

In this projection the eye is supposed to be at some
part of the earth's surface, say, upon the equator,
and it would in such case see the meridians of the
further hemisphere, as they would appear if produced
on to a flat surface tangential to the opposite point of
the equator which would be exactly under the eye. The
following (Fig. 7) will best illustrate this; in which we
suppose the globe to be transparent, and resting on a
point of its equator upon a table; the eye being at a,
and viewing the meridians, $c\,5$, $c\,4$, &c., as they would
appear to it if their intersection of the equator, $a\,b\,D$,
were produced to the surface of the table.

In this projection we require the use of the scale of chords, semi-tangents, tangents, sines, and secants. The visual rays from *a* passing and cutting the radius *b c*, form on it what is called the scale of semi-tangents; and the two triangles, *a c b* and "*a* D B," being *similar*, the divisions

Fig. 7.

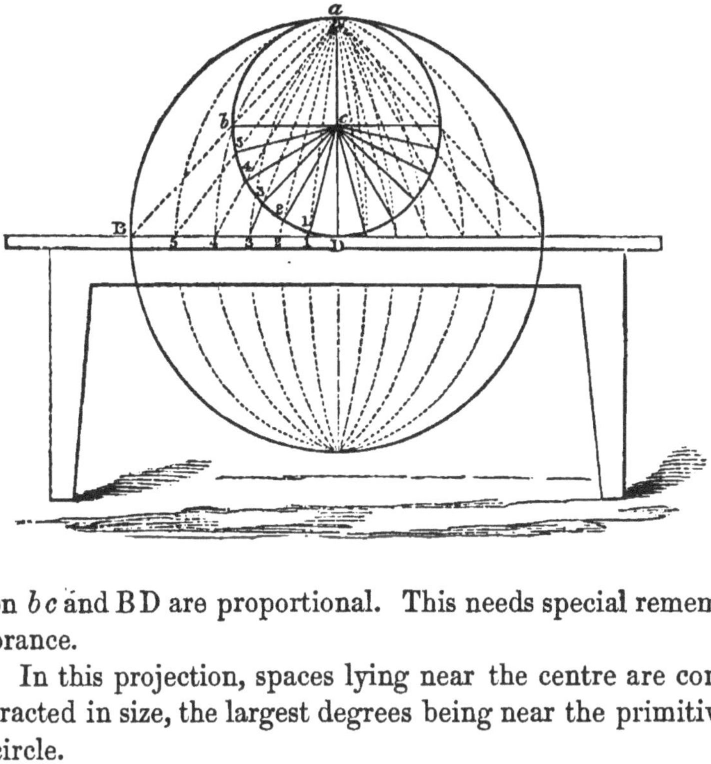

on *b c* and B D are proportional. This needs special remembrance.

In this projection, spaces lying near the centre are contracted in size, the largest degrees being near the primitive circle.

Meridians, as drawn obliquely, may be imitated by holding a ring or coin in various positions, the circumference at one time appearing as a circle, or at others as a straight line or slightly curved, &c.

GENERAL REMARKS ON
TRIGONOMETRY.

11. A few elementary hints, which, although essential to the proper understanding of the subject, are not always sufficiently explained to a beginner, and thereby materially retard his progress, will properly precede the consideration of spheric construction.

The following (Fig. 8) is called the trigonometrical " canon " (a word merely signifying a collection of mathematical truths or formulæ), and from it are derived all the terms and rules used and practised in trigonometry.

Fig. 8.

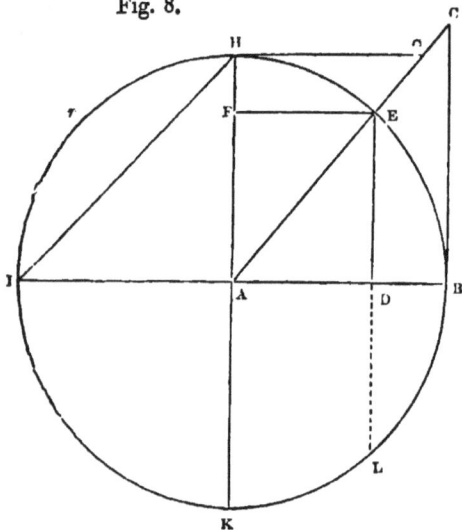

In the above,

AB is called the radius of the circle (of course A K, A I, A H, and A E, are also radii. Euc. I. def. 15.)

B C is called a tangent (always of the opposite angle A), and is so called from Latin, *tangere* to touch, because it only *touches* but does not cut the circle.

A C is called a secant [always of the angle between it

and radius, therefore of ∠ (angle) A], so called from Latin, *secare* to cut, because it cuts through the circle.

D E is called a sine, because it lies in the *hollow*, or *bosom*, of the curve E B L (Latin, *sinus*).

F E is called a cosine, or sine of an ∠ which is complementary to another, or required to make up 90°. Thus D E is the sine of the arc E B, and F E is the sine of arc E H, which is the complement of E B (for H E + E B = 90°). Therefore F E is called the cosine of E B, and is equal to A D, because F E and A D are drawn parallel, and E D is perpendicular to both.

G H is, in like manner, the tangent of H E, or cotangent of E B.

A G is, in like manner, the secant of H E, or cosecant of E B.

H I is called a line of chords, from its serving, as it were, to tie or confine the ends of the arc H *n* I.

D B is called a versed sine, and is the " height of the segment " E B L D E.

12. Every circle is supposed to be divided into 360 degrees (marked 360°). If, therefore, in the following figure, (9) P Z equals 90° or a quadrant, it is plain that if D *x* *also* equals 90°, the word " degree " refers to no measure of length, but merely signifies the 360th part of a circle, whatever the size of that circle may be; and, therefore, a degree may be of *any* length. As, however, degrees enter into calculations, *some definite value* of them *must* evidently be necessary; and, consequently, geometers express the *value* of degrees by taking any two lines from those given in the trigo-

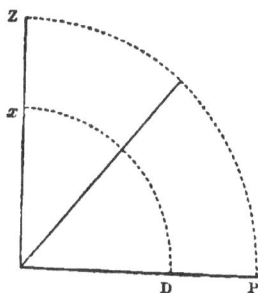

Fig. 9.

nometrical canon (Fig. 8), and consider the *length* of one *as compared* with the *length* of another in the same triangle: so that we use the terms sine, tangent, secant, &c., as referred generally to the radius of the circle, considering the length of radius to be 1 or 10, &c. (inches, feet, miles, leagues, &c., *at will*): this will be further illustrated.

13. It has long been customary to call a *line,* as E D (Fig. 8), a sine, or as B C, a tangent; but such is only correct when the length of a radius of a circle is known or understood. It is generally useful to describe the sine, &c., as fractions, thus, $\dfrac{ED}{AE}$ (Fig. 8), as they express fairly the value referred to. Anticipating by a few pages the question of proportion, it may here be noted that a vulgar fraction is a " ratio " or proportion in itself, and is deduced from a triangle. Thus, when we speak of $\frac{3}{4}$ths of anything we refer to some magnitude which can only be appreciated by considering the fraction $\frac{3}{4}$ in relation to its integer or whole number " one." As this whole number itself, expressed as a fraction, is $\frac{4}{4}$ths, $\frac{5}{5}$ths, $\frac{6}{6}$ths, &c. ; and when, therefore, we speak of $\frac{3}{4}$ths, we express a " ratio," meaning as 3 is to 4 (or symbolically 3 : 4), so then we express the value of

Fig. 10.

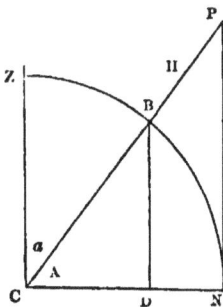

degrees by using ratios, and comparing them with the radius of the circle, which it is convenient to do by a number capable of decimal division, for obvious reasons (such as 1, 10, 100, &c.); and as at least one side of a plane triangle is always given, we are at liberty to compare this with the length of what is thus called the sine, tangent, secant, &c., of an angle, and hence the length of the arc itself. .

14. For further example, in Fig. 10, let B D be what is

commonly called the *sine* of the angle A. We describe its value by saying it is as the perpendicular is to radius, and write it thus: $\dfrac{\text{perp}}{\text{rad}}$ or from the figure $\dfrac{BD}{CB}$.

In like manner the other fundamental trigonometrical ratios are represented by fractions thus:—

$\dfrac{\text{perp}}{\text{rad}} = \dfrac{PN}{CN}$ is an expression for the tangent of $\angle$ A

$\dfrac{\text{hyp}}{\text{rad}} = \dfrac{CP}{CN}$ „   secant of $\angle$ A

$\dfrac{\text{rad}}{\text{perp}} = \dfrac{CN}{PN}$ „   cotang $\angle$ A

$\dfrac{\text{hyp}}{\text{rad}} = \dfrac{CP}{PN}$ „   cosec $\angle$ A

$\dfrac{\text{rad}}{\text{hyp}} = \dfrac{CN}{CP}$ „   cosine $\angle$ A

From the above it will be seen that certain ratios are *reciprocals;* for instance:—

$$\text{sine} = \frac{PN}{CP} \quad \text{and cosec} = \frac{CP}{PN}$$

$$\text{tang} = \frac{PN}{CN} \quad \text{and cotang} = \frac{CN}{PN}$$

$$\text{sec} = \frac{CP}{CN} \quad \text{and cosine} = \frac{CN}{CP}$$

15. Therefore, in works on logarithms, when we want the secant of an angle we can find it by subtracting its cosine from 20 (the diameter of a circle whose radius is 10), and to find the log sine we subtract the log cosec from 20; and to find the log tang we subtract the log cotang from 20, &c. Other useful deductions may be made, such as to find the log tang: add 10 to the log sine, and from the sum subtract the log cos (or log tang $= \dfrac{\log \text{sine} + 10)}{\log \text{cos}}$;

and to find the log cotang add 10 to the log cos, and from the sum subtract the log sine (or log cotang $= \dfrac{\log \text{cos} + 10)}{\log \text{sine}}$,

&c.; so that the values of the six fundamental ratios may be expressed thus :—

$$\text{sine} = \frac{1}{\text{cosec}} \qquad \text{cosine} = \frac{1}{\text{sec}}$$

$$\text{tang} = \frac{1}{\text{cot}} \qquad \text{cot} = \frac{1}{\text{tang}}$$

$$\text{sec} = \frac{1}{\text{cosine}} \qquad \text{cosec} = \frac{1}{\text{sine}}$$

N.B.—The unit here meaning 1 diameter $=2$ radii each of 10, or diameter $=20$.

16. This, however, which forms the elementary base of a proper knowledge of trigonometry, is not *absolutely essential* to the navigator, whose practical operations in plane trigonometry may be performed in total ignorance of principles, by dint of mere intelligence and skill from repetition; but in like manner does the blacksmith strike with the face of the hammer, and not with the handle, and would not probably perform his work more effectually if he were, previous to every blow, to calculate the force required to fashion his heated iron. But this must be remembered: a mathematical "blacksmith" would probably give *fewer blows*, because he would better know how to make each stroke tell, *from bringing the face of the hammer to bear on the iron in the best direction* with the greatest effect. In like manner the mathematical navigator will obtain his result in the shortest method.

It is beneath the dignity of a British sea officer to be content with mere knowledge of the use of formulæ. After reading this little book he may be safely advised to take up

" Jeans's Trigonometry," or some such *small work* on the subject.

DEFINITION OF AN ANGLE.

17. Before proceeding it may be well to explain what is really meant by an angle : that such explanation is necessary cannot be denied. A work of this description, which is designed as a mere stepping-stone to study, must needs adopt assertions without proofs, for fear of alarming the timid who *desire* improvement, but who yet doubt their own powers. It may, however, be safely asserted, that since our proofs are deduced mainly from the Books of Euclid a knowledge of his system of proving should be imparted at the earliest opportunity. In works upon navigation generally more extracts from Euclid are given than the sea officer thinks necessary for his satisfactory working, and too few to satisfy his after-desire of research; while the Books of Euclid themselves are supposed to be too heavy an undertaking for any but a schoolboy having no other employment than study. These are delusions. A groundwork in mathematics well laid is a continual source of mental profit and *amusement.* There is no limit (but the powers of mortal intellects) to the structure which may be raised upon it. A very long acquaintance with the subject of teaching can only lead to a belief that whenever mathematical study is to any mind found to be repulsive, it may be suspected that the individual student *has not had its details sufficiently explained.* Thousands upon thousands can work an equation by logarithms who have but an indistinct notion of what is really meant by an angle; and it may cheer the student when he sees that we may go even to the " dreaded " Euclid to obtain a full and clear comprehension even of this trifle.

For instance, he says among his definitions (Book I. def.

8): " A plane angle is the inclination of two lines to one another in a plane, which meet together, but are not in the same direction." And (Book I. def. 9), "A plane rectilineal angle is the inclination of two right (or straight) lines to one another, which meet together, but are not in the same right line ;" so that, in the following figure the right

Fig. 11.

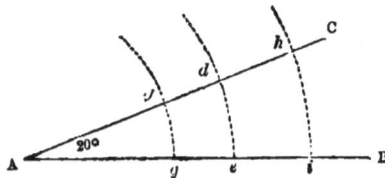

line A B meets the right line A C at the point A, and the " *angle* " is the inclination of these two lines as measured in degrees upon *any* circle drawn from A as a centre cutting these two lines. For instance, *de*, or *fg*, or *hi*, is each the measure of the " angle A " in degrees, 20 degrees meaning $\frac{20}{360}$ of *any* circle drawn round the point A.

GEOMETRICAL THEOREMS.

18. A few of Euclid's theorems may here be introduced with advantage.

Book I. XIII.—The angles which one right line makes with another upon one side of it, are either two right angles, or are together equal to two right angles. Departing from the precise language of absolute and complete proof (for obvious reasons), we may say that the semicircle H E B contains 180 degrees. If E G be perpendicular to A B, the arcs H E and E B being equal will each contain 90°, but B C is less than 90°, being, say, 55°; then E C must be 35°, and E H being 90°, C H will be 125°. Now, E C is called the *complement* of C B, and H C is called the *supplement* of C B.

Fig. 12.

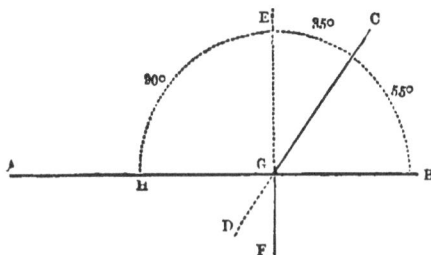

19. Book I. XV. tells us that if two right lines cut one another the vertical or opposite angles shall be equal. Thus, the angles C E A and B E D are equal to each other, as are also C E B and A E D; the angle C E A means the angle at the point E formed by the two lines C E E A. (We always put the letter *indicating the point* between the others.) Now, from the XIIIth Proposition, Book I., as above, it is evident that the two contiguous angles H G C and C G B equal 180 degrees; so in Fig. 13 $ac + cb$ equal 180 degrees, and $bd + da$ form the other 180 degrees.

Fig. 13.

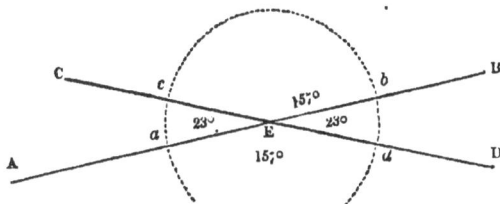

20. Euc. I. Prop. XXIX.—If a right line falls upon two parallel right lines, it makes the alternate angles equal to one another, &c. &c.; so that, indeed, A G E, C H E, F H D,

C

FGB, are equal to each other, being in this case about
28 degrees, while EGB, EHD, FHC, and FGA are also
equal, being about 152°,—the circles render this apparent.

Fig. 14.

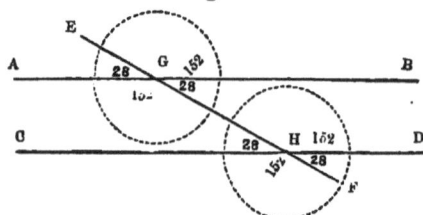

To satisfy this it needs only that AB and CD be precisely parallel.

21. Euc. I. Prop. XIX.—The greater angle of every
triangle is subtended by the greater side, or has the greater
side opposite to it.

The arcs of the circles drawn about each of the angular

Fig. 15.

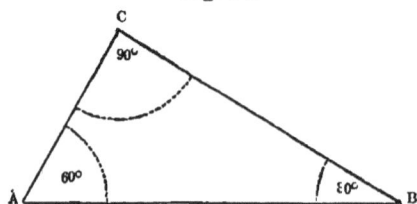

points with an equal radius show at once that, for instance, the small angle B 30° is opposite to the smaller
side AC, &c.

22. Euc. VI. Prop. VIII.—In a right-angled triangle
if a perpendicular be drawn from the right angle to the
base, the triangles on each side of it are similar to the
whole triangle and to one another (that is, triangle ADC,
triangle BDC, and triangle ACB are similar).

Now, by *similar* triangles we mean triangles which have

the three angles in the one equal to three angles respectively in the other; and although their opposite sides may be of different lengths, they are nevertheless proportional: thus, by the figure, if we lay off the distance C D at B*c*, and draw *a c* parallel to C D, we shall find the triangle B*ca* equal in its angles to C D A, and its sides proportional to triangle B C D, that is, B*c* will be to *ac* as B D is to D C (B*c* : *ac* :: B D : D C), &c. &c., and here (by I., XXIX.), because B C falls across the two parallel lines *ac*, C D, the angles B*ac* and B C D are equal. (And this is the

Fig. 16.

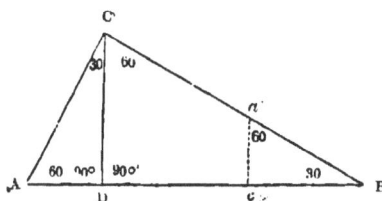

way in which one proposition of Euclid rests for proof upon others which precede. As an example, proof of the above proposition could only be made with *mathematical accuracy* by reference to 34 propositions of Book I., 10 of Book V., and 3 of Book VI.; in all 47 *propositions*, besides *definitions, axioms, and postulates*, repetitions, &c.).

23. Euc. III. Prop. XX. — The angle at the centre of a circle is double the angle at the circumference upon the same base, that is, upon the same part of the circumference.

In the triangle A D C, the side A D equals D C, therefore, as equal sides are opposite to equal angles (as deduced from Book I. Prop. XVIII.), the angle D A C equals D C A. But Prop. XXXII. Book I. implies that the angle E D A equals ‚D A C and D C A together; let A C B and

A D B be two angles standing upon the same base A B; therefore, as A D E is the double of A C D, and by like reasoning E D B would be the double of E C B, so must the whole angle A D B be the double of the angle A C B.

Fig. 17.

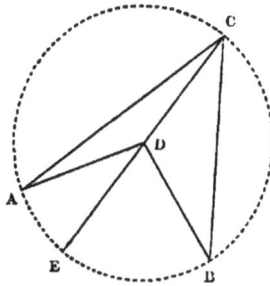

24. Euc. I. Prop. XXXII.—If any side of a triangle be produced, the exterior angle is equal to the two interior and opposite angles; and the three interior angles of any triangle are equal to two right angles.

Fig. 18.

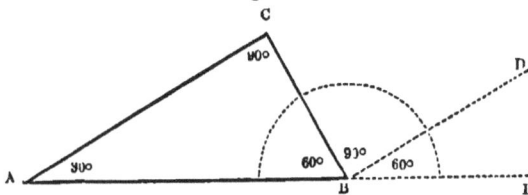

Reference to Prop. XXIX. Book I. will show that the angle C A B will equal the angle D B E, because the right line C B falls on the two parallels A C and B D; also, that the angle A C B will equal C B D. Therefore the angle C B E, which is exterior to the triangle, and is made up of the two angles D B E and D B C, is equal to the two interior and *opposite* angles of the triangle (the angle A B C being the *adjacent* angle). And it is also evident that the two "opposite" angles, *together with* the "adjacent" angle form the *interior* angles of the triangle) are equal to the exterior and adjacent angles together, therefore (by I. XIII.) are equal to 180° or two right angles.

The above will be sufficient to give a general notion of the importance of Euclid.

25. Every triangle has three sides, and in lettering a right-angled triangle it is customary to place B at the right angle and to make A C designate the hypothenuse.

RATIOS.

26. The rules of Plane Trigonometry do not fall within the present compass of this work; they are to be found in all works on navigation. It may, however, be remarked that the rules given therein for the working of questions in navigation, take for instance—

> as diff lat
> is to rad
> so is dep
> to tang course,

are in the form of a proportion or ratio (proportion is the equality of ratios)—three things, as in what is called the Rule of Three, being given to find a fourth. Every conceivable arithmetical calculation is the working of a proportion, or the comparison of ratios. When we say 5 times 8 make 40, we mean to say that $1 : 5 :: 8 : 40$, or, fractionally, $\frac{1}{5}$, $\frac{8}{40}$; and these fractions form what is called an "equation," for $\frac{1}{5} = \frac{8}{40}$. But Euclid, Book VI. 16, demonstrates that, if four straight lines (or quantities) be proportionals, the product of the means (or middle terms) of such proportion shall be equal to the product of the extremes (or first and last terms), as from the above $1 \times 40 = 5 \times 8$; and, as another example, if 3 coils of rope cost 20 shillings, 6 coils will cost 40, or $3 : 20 :: 6 : 40$; that is $3 \times 40 = 20 \times 6$.

27. Euclid, further, in Book V. (Def. 13 to 17), shows that such proportions may be varied by division, conversion, inversion, alternation, &c., so that the last example admits of being varied in form; for

As 20 : 40 :: 3 : 6

6 : 40 :: 3 : 20

40 : 6 :: 20 : 3, &c.

So far, therefore, as the calculation of straight lines is concerned (which may be put to represent by their proportionate *lengths* any proportionate *quantities*), the work of calculation becomes easy; but in entering upon the calculation of angles, 'it should be shown that common arithmetic fails altogether in its powers to readily solve all the parts of a trigonometrical figure. Nor is the difficulty properly explained in elementary works of the present day.

NATURAL SINES.

28. If we examine Fig. 19, in which the circumference is divided into equal parts of ten degrees each, we shall see in the line C D a scale of what are called "Natural Sines." C 30 being equal to *a* 30, &c. The divisions on

Fig. 19.

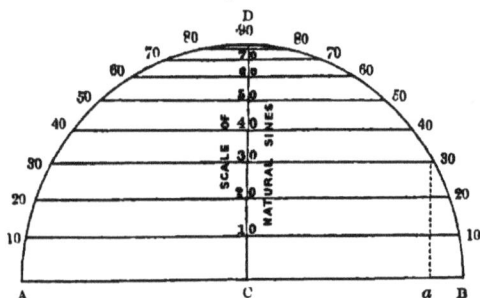

C D, moreover, are not equal; for C 10 is much larger than the distance 40 to 50; and, indeed, the sine of 30° is exactly in length half the sine of 90°, while the natural number 30° is only one third of 90°; and we shall also find that the secant of 60° is equal to twice the sine of 90°, while the natural number 60 is two thirds of 90; that the tangent of 45° is equal to the sine of 90°, &c.

LOGARITHMS.

29. To obviate the above inconvenience, an *artificial* system of numbers was sought for by mathematicians; and, happily, Lord Napier, about 250 years since, gave the world his System of Logarithms; and our astonishment is excited when we learn that this great discovery was made nearly half a century before what has since been called the "Logarithmic Series" was invented.

NATURE OF LOGARITHMS.

30. The very word *Logarithm* is a stumbling-block to many; but it is easy to show that, although the ground-work is so very little known to many who use logarithms, a few hints put in familiar language will not merely gratify a laudable curiosity, but pleasantly assist in further investigation.

31. Lord Napier based his system of logarithms upon the following infinite series, in which it will be seen that values of fractions are systematically diminished by adding increasing multipliers to the denominators.

$$1 + \frac{1}{1} + \frac{1}{1 \cdot 2} + \frac{1}{1 \cdot 2 \cdot 3} + \frac{1}{1 \cdot 2 \cdot 3 \cdot 4} + \frac{1}{1 \cdot 2 \cdot 3 \cdot 4 \cdot 5} + \&c.$$

Any one who understands vulgar fractions can resolve these into the following:—

$$1 + 1 + \tfrac{1}{2} + \tfrac{1}{6} + \tfrac{1}{24} + \tfrac{1}{120} + \&c.$$

Of course the series might be extended to great length

and increased accuracy, but the above is enough for our purpose. Now, if we reduce the above fractions to a common denominator, we get

$$2 + \tfrac{60}{120} + \tfrac{20}{120} + \tfrac{5}{120} + \tfrac{1}{120}, \&c.;$$

and adding numerators we get $2\tfrac{86}{120}$, which, reduced to decimals, becomes $= 2 \cdot 71666$, an *approximate* base of the Napierian logarithms, which, if extended to further terms, and the division be made as usually shown in works on logarithms, becomes (as under) $2 \cdot 7182818$, &c., which, so far as it extends, is the *true* base of the Napierian logarithmic system; thus :—

$$1 + 1 + \tfrac{1}{1 \cdot 2} \qquad = \quad 2\tfrac{1}{2} \quad = 2 \cdot 5$$

$$\frac{1}{1 \cdot 2 \cdot 3} \qquad = \quad \frac{1}{6} \quad = \cdot 1666666666$$

$$\frac{1}{1 \cdot 2 \cdot 3 \cdot 4} \qquad = \quad \frac{1}{24} \quad = \cdot 0416666666$$

$$\frac{1}{1 \cdot 2 \cdot 3 \cdot 4 \cdot 5} \qquad = \quad \frac{1}{120} \quad = \cdot 0083333333$$

$$\frac{1}{1 \cdot 2 \cdot 3 \cdot 4 \cdot 5 \cdot 6} \qquad = \quad \frac{1}{720} \quad = \cdot 0013888888$$

$$\frac{1}{1 \cdot 2 \cdot 3 \cdot 4 \cdot 5 \cdot 6 \cdot 7} \qquad = \quad \frac{1}{5040} \quad = \cdot 0001984126$$

$$\frac{1}{1 \cdot 2 \cdot 3 \cdot 4 \cdot 5 \cdot 6 \cdot 7 \cdot 8} \qquad = \quad \frac{1}{40320} = \cdot 0000248015$$

$$\frac{1}{1 \cdot 2 \cdot 3 \cdot 4 \cdot 5 \cdot 6 \cdot 7 \cdot 8 \cdot 9} \qquad = \quad \frac{1}{362880} = \cdot 0000027557$$

$$\frac{1}{1 \cdot 2 \cdot 3 \cdot 4 \cdot 5 \cdot 6 \cdot 7 \cdot 8 \cdot 9 \cdot 10} = \frac{1}{3628800} = \cdot 0000002755$$

$$\overline{\qquad 2 \cdot 7182818 \qquad \&c.}$$

Now, the Napierian system, arising, as above shown, from

so simple an infinite series, and one which is so easily remembered, is made the basis of all other systems.

32. Any number may be taken as a base; let us at
random take the number 3. Then, as logarithms are de-
fined to be *a series of numbers in arithmetical progression, placed opposite to and corresponding with another
series in geometrical progression, and so placed that* 0
in the logarithmic stands opposite 1 *in the geometric*—we
can easily form a skeleton system based on the number 3,
as under:—

Natural Numbers.	Geometrical.		Logarithms.
1 =	1	. . . .	0·000000
3^1 =	3	. . . .	1·000000
3^2 =	9 (3 × 3)	. . .	2·000000
3^3 =	27 (3 × 3 × 3)	. .	3·000000
3^4 =	81 (3 × 3 × 3 × 3)	.	4·000000
3^5 =	243 &c. &c.	. .	5·000000
3^6 =	729	. . . .	6·000000
3^7 =	2187	. . . .	7·000000
3^8 =	6561	. . . .	8·000000
3^9 =	19683	. . . .	9·000000
3^{10} =	59049	. . . .	10·000000

33. To prove that the above is really a table of logarithms, let us attempt calculation by it as we would by the
table in common use: remembering the rules, that—

In logarithms we *multiply* numbers by *adding* their
logarithms, and we *divide* numbers by *subtracting* their
logarithms. Suppose, for example, we desire to multiply
729 by 81.

In the above table
the log of 729 is　6
　　log of　81 is　4
　　the sum　10 = log of the answer, 59049

By arithmetic.

$$729$$
$$81$$
$$\overline{729}$$
$$5832$$
$$\overline{59049}$$

And again, to divide 2187 by 243,

By arithmetic.	By logarithms.
243)2187(9	log of 2187 is 7
2187	log of 243 is 5
	the difference $\overline{2}$ = log of 9.

And further, to extract the cube root of 19683. This is done by dividing the logarithm of the number by the index of the power.

$$\log \text{ of } 19683 = 9, \text{and } \frac{9}{3} \text{ the index} = 3 \text{ which is the log of 27;}$$

or, as written $\sqrt[3]{19683} = 27$

and again, to raise the number 9 to the fourth power: multiply the log by the number of the index of the power; thus—

$$\log \text{ of } 9 \text{ is } \quad 2$$
$$\text{index } 4$$
$$\overline{8} = \log \text{ of the number } 6561;$$

or $9^4 = 6561.$

34. A table of the above description has, however, serious defects; the greatest is apparently the want of analogy between the number of figures in the whole number and the index of the logarithm, as will be shown immediately. The index or *characteristic* of the logarithm, is the integer, or whole number, and the decimal is called the *mantissa*. To facilitate calculation by logarithms, Mr. Henry Briggs, a contemporary of Lord Napier, published at Cambridge, in 1615, a system having for its base the number 10, the root of our decimal scale of notation, in which the powers of the number 10 are shown by merely adding to unity as many figures as are " indicated," by what are therefore aptly called the indices of different powers, as we see in the following skeleton table to the base 10. Here, again, the logarithm is merely the *index of the power*, while it indicates absolutely the number of figures, *less one*, in the whole number to which it corresponds. This is the common system of logarithms.

35. The above table (32) only gives the logarithms of numbers which are multiples or powers of 10, but we might, for instance, require to know the log of 270 — which would evidently lie between the log of 100 and the log of 1000; its index, however, would be 2 (because the number contains three figures), together with a decimal or mantissa, and we see in a more extended table it would be as represented by the log 2·23044.

Natural Numbers.		Logarithms.
$10^0 = 1$	· · · ·	0·000000 &c.
$10^1 = 10$	· · · ·	·1·000000
$10^2 = 100$	· · · ·	2·000000
$10^3 = 1000$	· · · ·	3·000000
$10^4 = 10000$	· · · ·	4·000000
$10^5 = 100000$	· · ·	5·000000
$10^6 = 1000000$	· · ·	6·000000
$10^7 = 10000000$	· · ·	7·000000
$10^8 = 100000000$	· · ·	8·000000
$10^9 = 1000000000$	· ·	9·000000
$10^{10} = 10000000000$	· ·	10·000000

36. It is important to add, that if the natural number be a vulgar fraction, such as ⅝, we may (because it means 5 divided by 8), subtract the log of the denominator from that of the numerator (increased by unity if necessary) thus—

$$\begin{array}{ll} \text{log of } 5 = 0 \cdot 698970 \\ \text{,, ,, } 8 = 0 \cdot 903090 \end{array} \quad \text{Proof} \quad \begin{array}{r} 8)5000 \\ \hline \cdot 625 \end{array}$$

$$\overline{1} \cdot 795880 = \text{dec. fraction } \cdot 625.$$

It is obvious, therefore, that it would have been as simple to have reduced the vulgar fraction to its decimal at once, and then taken its logarithm.

37. But we borrowed an unit in subtracting, therefore the resulting 9 was absolutely ·9 (decimal 9) or minus 1 (written 1). (A word here to those who are not "well up" in decimal arithmetic; be advised and lose not a day in "brushing up" a little. It is not, however, likely, that any one having sufficient interest in the subject to enable him to read this little book thus far, will do otherwise.) It will then be easily seen that it would, in the above example, have been better to borrow 10 than 1, writing the resulting index 9 as *minus* 9 (− 9). Another example : multiply 100·6 by ·1006.

$$\begin{array}{llr} \text{log of } 100 \cdot 6 & = & 2 \cdot 002598 \\ \text{,, } \quad\quad \cdot 1006 & = & -9 \cdot 002598 \\ \hline & & 1 \cdot 005196 \end{array}$$

casting off the borrowed ten it will be 1·005196, equal to the number 10·12; this is a more simple plan than writing the log of ·1006, as $\overline{1}$·002598, and operating algebraically.

The following abstract will have its uses, and illustrate the above. (The number 1006 is taken at random, any number may be substituted.)

Natural Numbers.						Logarithms.
1006	.	.	.	.	.	. 3·002598
100·6	.	.	.	.	.	. 2·002598
10·06	.	.	.	.	.	. 1·002598
1·006	.	.	.	.	.	. 0·002598
·1006	.	.	.	.		—9·002598
·01006	.	.	.	.		—8·002528
·001006	.	.	.	.		—7·002598
&c.						

N.B.—All works on logarithms have rules attached, for taking out numbers, whether representing linear or angular quantities.

COMPUTATION OF LOGARITHMS.

38. We have seen (34) that only logarithms which have a certain base are conveniently applicable to practical pur poses, and that the Napierian system is the most simply obtained from a *series*, which gives its base 2·718281828, &c. This is commonly called the *natural* or hyperbolic system, and is written thus—

Log$_\varepsilon$ 2·718281828, &c., while the decimal or Brigg's or the common system is written log $_{10}$ (read, log to the base 10, or log to the base ε).

We use the Napierian system as a foundation of our common system in the following deduced formula :—

$$\left.\begin{array}{l}\text{The common log}\\ \text{of any number}\end{array}\right\} = \frac{\text{Nap. log of number}}{\text{the Nap. log of 10}} = \frac{\text{Nap. log. of number}}{2\cdot3025851} =$$

$$\frac{1}{2\cdot3025851} = \cdot43429448 \times \text{Nap. log of the number.}$$

Hence, to construct the common logarithm of any number, we use a number which may be called a *Napierian constant ;* it is the double of the above equals ·43429448 (which is the *modulus* of the common system), and equals ·86858896.

It is here quite unnecesary to give the algebraic reasons why we use the following infinite series: it is enough for our purpose to say that in it we have a series, by which we may compute the logs of all natural numbers, and this without knowing the log of any previous number. It is this:—

$$\log P = 2M \left\{ \frac{P-1}{P+1} + \tfrac{1}{3}\left(\frac{P-1}{P+1}\right)^3 + \tfrac{1}{5}\left(\frac{P-1}{P+1}\right)^5 + \tfrac{1}{7}\left(\frac{P-1}{P+1}\right)^7 + \&c. \right.$$

Now, if we let P represent the number whose log we require, say the number 2, and M the modulus, ·43429448, the above will, in figures, be as under:—

$$\log 2 = 2(\cdot 43429448)\left\{ \frac{2-1}{2+1} + \frac{1}{3}\left(\frac{2-1}{2+1}\right)^3 + \frac{1}{5}\left(\frac{2-1}{2+1}\right)^5 + \frac{1}{7}\left(\frac{2-1}{2+1}\right)^7 + \&c. \right\}$$

$$\log 2 = \cdot 86858896 \left\{ \frac{1}{3} + \frac{1}{3}\left(\frac{1}{3}\right)^3 + \frac{1}{5}\left(\frac{1}{3}\right)^5 + \frac{1}{7}\left(\frac{1}{3}\right)^7 + \&c. \right\}$$

(or, as reduced)

$$\log 2 = \cdot 86858896 \left(\frac{1}{3} + \frac{1}{81} + \frac{1}{1215} + \frac{1}{15309} + \&c. \right)$$

(or decimally)

$$\log 2 = \cdot 86858896 \;(\cdot 333333 + \cdot 012347 + \cdot 000823 + \cdot 000065 + \&c.)$$

(or, after addition)

$$\log 2 = \cdot 86858896 \times \cdot 346568$$

(or, after multiplication)

$\log 2 = \cdot 301025$ approximately, but if the series be extended

$\log 2 = \cdot 3010300$ as we find in our tables of common logarithms.

NAUTICAL ASTRONOMY.

39. Before entering upon precise rules for construction and calculation of spheric angles, a few remarks upon the subject of nautical astronomy itself will lighten the labour of the student.

The general complaint of those who have " looked into " spherics, is, that although they have been taught to work spheric angles, they do not understand the principles sufficiently to be able to apply their knowledge with confidence to ordinary or rather extraordinary work.

If, say they, we could always see the figures illustrating our questions, were it only in the mind's eye, it would assist us in obtaining solutions, and increase our interest in the study; it would also very materially help our memories.

Too generally, however, a spheric triangle is drawn by hand, without any reference whatever to its adaptation to .the data of the question under consideration; hence much theoretical difficulty arises from the consideration of angles, as acute or obtuse, and of complements, supplements, &c. If in the study of spherics, the maturing of the reasoning powers is to be our main object (as in teaching Euclid for the mere logic of its reasonings), present works on the subject are abundant. But if the study is to be undertaken by those circumstanced like sea officers in general, whose object is to master enough of the principles to enable them to pursue their professional avocations in mathematics, with confidence and success, an abridged arrangement of scientific facts suitable to their purpose will doubly benefit them, inasmuch as not only will their nautical work be more accurately and more readily per-
. formed, but the having once obtained a well-grounded acquirement in principles, will render it less difficult for

them to employ their few hours of leisure, which some
services permit, in pleasant advancement.

. **40.** The special object of the following explanations is
then, to lead the young seaman to the most agreeable
part of a navigator's study, viz., "construction." Not
that in practice he will be required to actually draw his
diagrams, *but a knowledge of construction will greatly
aid him in calculation.*

LINES OF THE SPHERE.

41. To give a general notion of the imaginary lines
of the sphere, we will suppose that I am standing on the
coast, looking seaward towards the west. The north would
in such case evidently be on my right hand, the south
on my left hand, and the east would be behind me. I
might imagine a point over my head called the zenith,
and a point below my feet called the nadir. The distance
of the zenith and nadir would, of course, be unlimited;
but let us for the sake of precise illustration limit it to
any distance, say, 1000 yards; in such case I must consider the horizon to the north and south also limited to
the same distance. Now, suppose further, the meridian
of the place on which I am standing to be a line drawn
from the north point of the horizon, up over my head
through the zenith point, and down precisely to the south
point of the horizon, this would give me a semicircle.
And again, imagine lines drawn from the point over my
head (zenith) downwards, so as to cut each point of the compass at the horizon; these would be called azimuth circles.

Let me now further suppose that I walk backwards in
a line due east, until I see the figure my imagination has
been constructing with, say, a radius of 1000 yards; it

would, if visible, appear precisely as the following figure, 20 (if drawn on the stereographic projection, which alone will be used, with a slight exception, in the following demonstration of spherics), the centre marked west (W). being the point on which I had been previously standing.

Fig. 20.

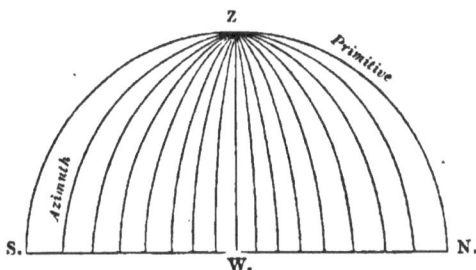

Suppose, again, that the centre of this figure was situated in latitude 50° N., and for illustration, that the polar star is exactly at the north polar point of the heavens, or exactly over the north pole of the earth (which it is not by about 1½ degrees), let me now, as in Fig. 21, divide my figure from north to Z, and from Z to south into degrees, 90 in each quadrant; and through the 50th degree from north or N. I place the pole of the heavens, being at about the spot at which I should see the polar star at night in the heavens, while standing at the centre of the figure. I would next imagine a line connecting this pole with the centre of the figure, and drawing another line at right angles to it from W to Q, the latter would represent the part of the equator above the horizon, S N, because every part of it would be 90° from the pole. It has already been shown that the great circle, N Z S, is a meridian; it is also to an observer at W, the 12 o'clock " *hour circle;* " because, supposing the sun to be in north declination (or distance north of the equator), and rising at the point of the horizon marked *c*, it would, between

D

its rising and noon, seem to describe the small arc, *c d*,
until, being at *d*, on the meridian of the place at noon,
it would descend from *d* towards *c*, where it would
" set " below the horizon. Thus we see that the point W
answers either for east or west. In Figs. 20 and 21, the
azimuths are drawn to every point of the compass,
and whichever azimuth circle cuts the horizon at the
point of the sun's setting would be its true bearing
at such sunset; in Fig. 21, it would set at about
W.N.W. But if P Z S is an hour circle, so P W would
be another (viz., the 6 o'clock hour circle), and we
may conceive others to be drawn intermediate. P *y* is
therefore one, and represents the hour circle of about half-
past 2 P.M. or ½ past 9 A.M., while O the intersection of *d c*
and P O *y*, would be the sun's place at that time; and it is,
for example, the work of " spherics " to calculate the proportions of the spheric triangle Z O P, P O being evidently
the " polar distance," and Z O the zenith distance, Z P the
co-latitude, &c.

Fig. 21.

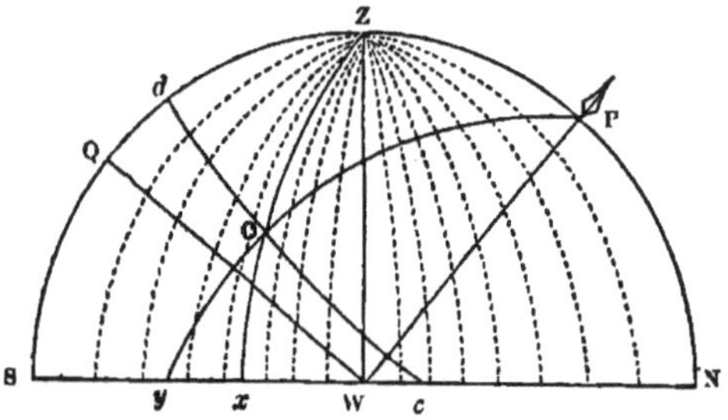

SPHERIC TRIANGLE.

42. We may now, by another figure, 22, show that
what is meant by a spheric triangle is really a part of the
surface of a solid globe, bounded by three arcs of great
circles, as the triangle A B C. The angles being the in-

clinations of the planes of the great circle to each other,
and the lengths of the sides have always reference to the
angles they make at the centre of the solid figure, although only the triangle itself is usually shown in diagrams.

Fig. 22.

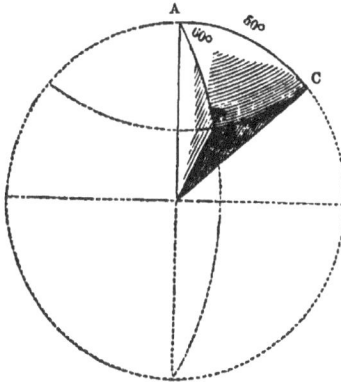

Having now, it is presumed, given a correct idea of
what is to be understood by "spherics," it will be pleasant to see how this very interesting branch of science is
illustrative of daily ordinary phenomena.   If we inquire as
to what extent of information might be obtained from very
simple data, the reply will be highly encouraging, because
we shall find that the drawing of a few curves by *very
easy* rules (which are about to be fully explained further
on) open to us a vast and satisfactory insight into astronomy itself.

GENERAL DIAGRAM OF DAILY ASTRONOMICAL PHENOMENA.

43. Let it be supposed that we are in latitude 50° N., that the date is the 20th May, and the time of day 10 in the morning.

N.B.—On the 20th May the sun's declination or distance from the equator would be about 20° N.

Figure 23 is drawn from these data, and gives us the following information accordingly :—

Fig. 23.

Data.—Lat 50° N =P N
Dec 20° N =d E or O I
Time 10 A.M.=∠Q P O

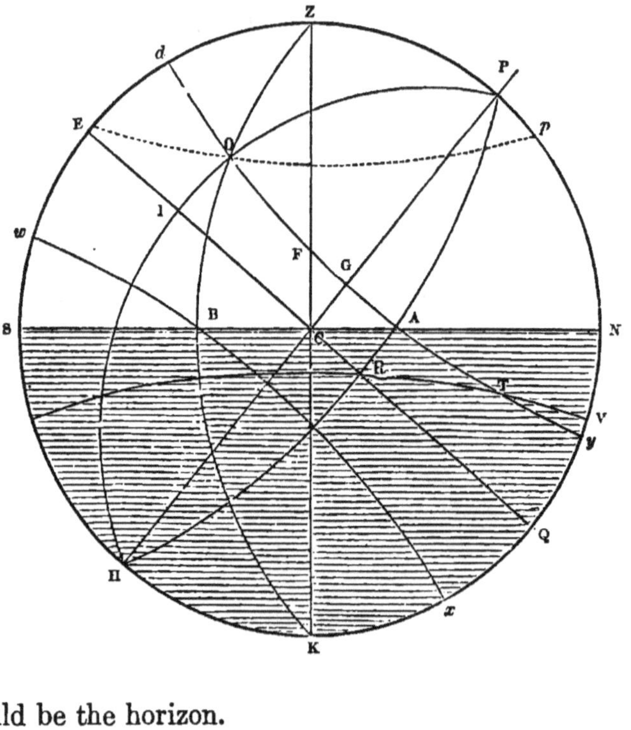

S N would be the horizon.
O B „ the sun's altitude.
O Z „ „ zenith distance.
O P „ „ polar distance.

E p would be a parallel of altitude.

Z	„	the zenith.
K	„	„ nadir.
N	„	„ north part of the horizon.
S	„	„ south.
C	„	„ east or west part of the horizon.
O d	„	„ sun's distance from the meridian.
Z O B	„	„ azimuth circle.
S B	„	„ azimuth from south.
B N	„	„ „ north.
C B	„	„ amplitude from C (east or west).
Z d	„	„ meridional zenith distance.
S d	„	„ altitude at noon (meridional altitude).
P G H	„	„ six o'clock, hour circle.
P O H	„	„ ten o'clock, or 2 P.M. hour circle.
Z C	„	„ prime vertical.
$d y$	„	„ declination, or apparent path in the heavens for the day.
d A	„	half the length of the day.
A y	„	„ „ night.
y	„	sun's place at midnight.
A	„	„ „ rising or setting.
d	„	„ „ noon.
O	„	„ „ 10 A.M., or 2 P.M.
G	„	„ „ 6 o'clock.
F	„	„ „ when on the prime vertical.
$\angle$ Q P R	„	time of the sun's rising or setting.
A T	„	the limit of duration of twilight.
d O y	„	parallel of declination, 20th May.
w B x	„	„ „ 21st December.
B w	„	half the length of the shortest winter's day.
B x	„	„ „ longest winter's night.
N x	„	the sun's distance below the horizon on 21st December at midnight.

S *w* would be the sun's meridian altitude on 21st December.

N P „ the latitude of theplace (or heightof the pole).

Z P „ complement of latitude.

A G „ ascensional difference.

SPHERIC PROJECTION.

44. The general terms used in nautical astronomy being thus understood, it remains to illustrate "spheric projection;" but as we mean to explain as we go, let us advance warily. Only those problems which are absolutely essential to sea officers will be at first given. The remainder may possibly follow in a supplementary volume for the assistance of those who desire a more extensive acquaintance with the subject.

In the following figures the eye is supposed to be opposite the centre, which point is called the " pole " of the primitive or boundary circle; but the word pole will not henceforward in this book signify anything more than a point exactly 90° from some great circle.

Circles are either great or small, not so much from their dimensions as from their *position on the sphere.* None but great circles can divide a sphere into two equal parts, their planes cutting the centre. Small circles are those whose planes do not cut the centre, but divide the sphere into two unequal parts. Small circles are also called parallel circles, because their planes are parallel to the plane of the equator (of this kind are parallels of latitude, of declination, of altitude, &c.).

Before explaining the problems, the following is to be specially remembered, viz., all the measures used in spherics, such as sines, tangents, &c., are taken from parts of the circle; and in case the student may have forgotten the construction of the plane scale, its formation may be

seen in the accompanying diagram, from which all measures in the succeeding figures are drawn.

45. PLANE SCALE.

Fig. 24.

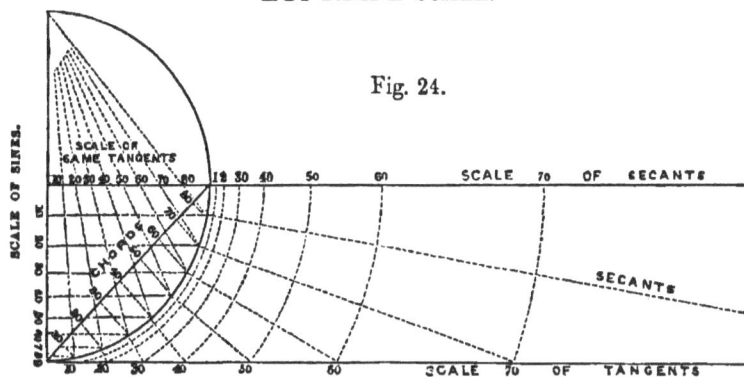

PROBLEMS.

46. PROB. I.—*To make an angle so that the angular point
shall be at the centre of the primitive (say 50°).*

From centre O through 50° on the primitive draw O *a*
if the arc be already divided. If not divided, make the
angle *a* O H equal 50° by the use of a scale of chords, laying 50° from H to *a* (first drawing the circle with a radius
of 60° from the scale, as in all cases).

Fig. 25.

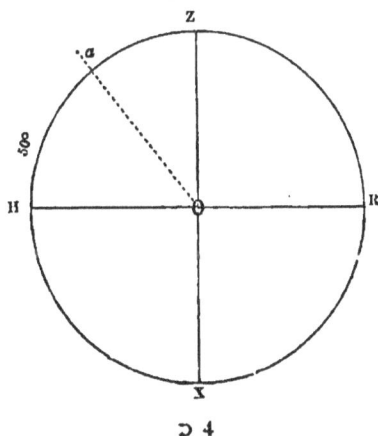

D 4

47. PROB. II.—*To lay off any number of degrees on a right circle (say 60°).*

Fig. 26.

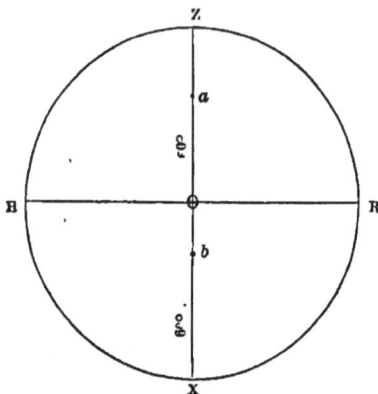

1st (say upon O Z from the centre). Take 60° from the scale of semi-tangents and lay off from the centre to *a*.

2nd (say upon X O from X towards O). On the scale of semi-tangents count 60° *backwards from* 90°, and lay it off from X towards O, at *b*.

48. PROB. III.—*To draw an oblique circle through any point lying within the primitive circle.*

Fig. 27.

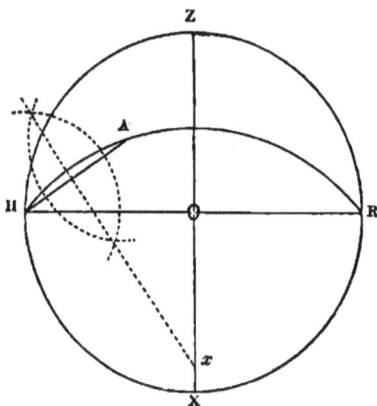

Connect the point (as at A) with the primitive H Z R X, and from the point (say H) at the primitive draw a diameter H O R ; draw also another diameter at right angles to the first, as Z O X. Bisect the line A H, and produce (or lengthen) the bisecting line till it cuts O X at x; then x will be the centre of an oblique great circle which will cut the point A.

49. PROB. IV.—*To draw an oblique circle through any two points, say through point I and point A.*

Through either point (say A) draw a diameter as H A O R· Draw Z X at right angles to it at the centre. Join A Z. Make A Zy a right angle at Z. The intersection of Zy on

Fig. 28.

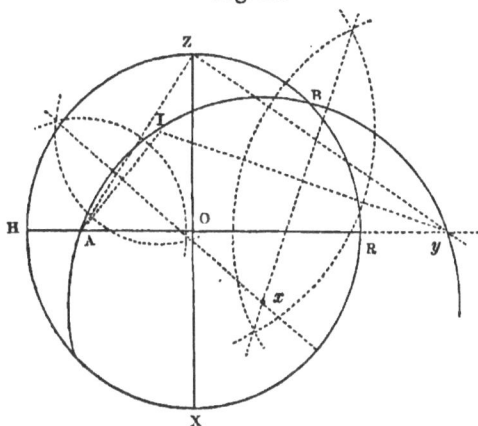

H R produced will give a third point, y. Join Iy and A I, and bisect Zy and A I, the lines of intersection will meet at x, which will be the centre of the great circle A I B.

50. PROB. V.—*To draw a small circle parallel to the primitive at any distance from it (say 40°).*

With its complement 50° from the scale of semi-tangents

and centre O, draw a circle as *a b c*, and it will be 40° from the primitive.

Fig. 29.

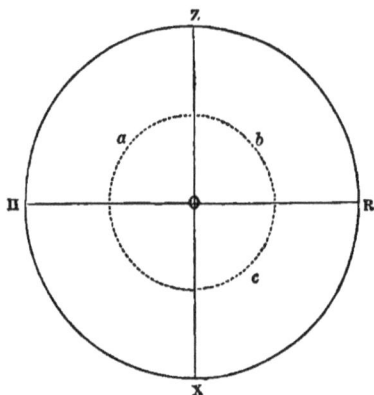

51. Prob. VI.—*To draw a parallel circle at any given distance from a right circle (say at 50° from H R), or at any distance about a given point at the primitive (say at 40° from Z.)*

1st. Lay off 50° from the scale of chords from H to *a*,

Fig. 30.

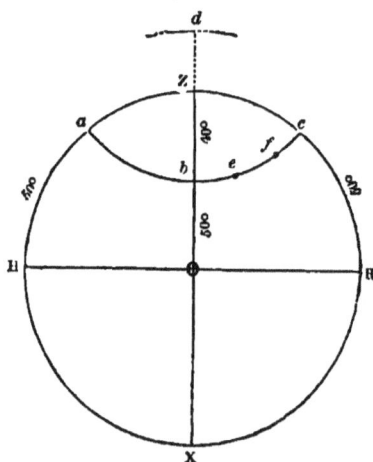

also from R to *c*, and 50° from the scale of semi-tangents from the centre to *b*. Then through the three points *abc* describe a circle as in Prob. IV. Or, take the secant of (90−50°)=40°. Lay off this distance from O to *d*, and then, with the tangent of 40° and *d* as a centre, draw *abc*, the small circle required. Whether the point be *b*, or *e*, or *f*, &c., use the same means.

52. PROB. VII.—*To find the pole of an oblique circle (say of ZcX).*

Draw and produce X*c* till it cuts the primitive. From *a* lay off the distance of 90° (or HZ) from *a* past Z to *b*.

Fig. 31.

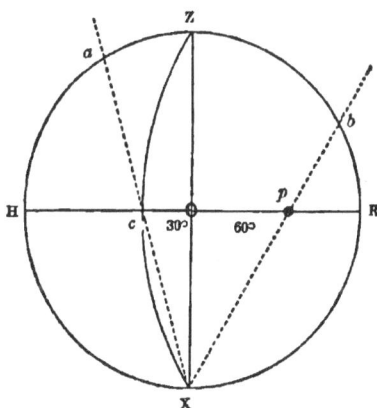

Join *b*X; and where this line cuts the radius OR (or at *p*) will be the pole of ZcX. Or thus: Measure CO on the scale of semitangents, and lay off its complement from O towards R, as at *p*, the pole.

53. PROB. VIII.—*To draw a great circle through any given point so as to make any desired angle at the primitive (say 40° and through the point d).*

With centre O and tangent 40° describe an arc as *ab*;

with centre d and secant 40° describe another arc which
cuts the first at x; then x is the centre of the oblique
circle ZdX, and it is drawn accordingly.

Fig. 32.

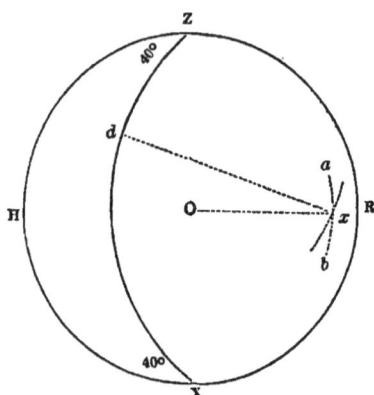

Note.—Diameters are always at right angles to the
primitive.

54. PROB. IX.—*To draw a* small *circle through a given
point* (x), *which shall be at a given distance from a
right circle* (*say at* 40°).

Take the secant of 50° (the complement of the given

Fig. 33.

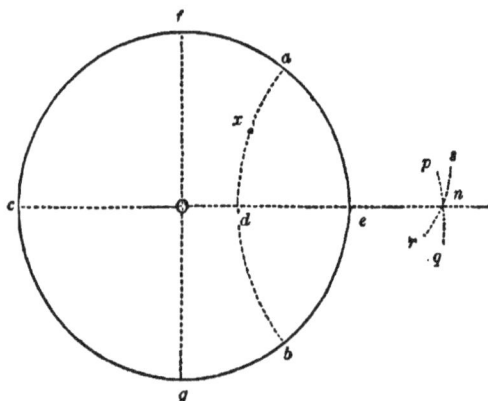

distance) and from centre O sweep an arc as pq; cross this with the tangent of 50° laid off from the point x. Then will n be the centre of the small circle adb. Join On, and draw the diameters ce and fg. Then gb and fa will each measure 40° on the scale of chords, and Od will measure 40° on the scale of semi-tangents.

55. PROB. X.—*To draw an oblique circle perpendicular to a given oblique circle.*

Find the pole p of the given circle ZaX (Prob. VIII.). Draw any diameter at pleasure, say bOc; through cpb draw

Fig. 34.

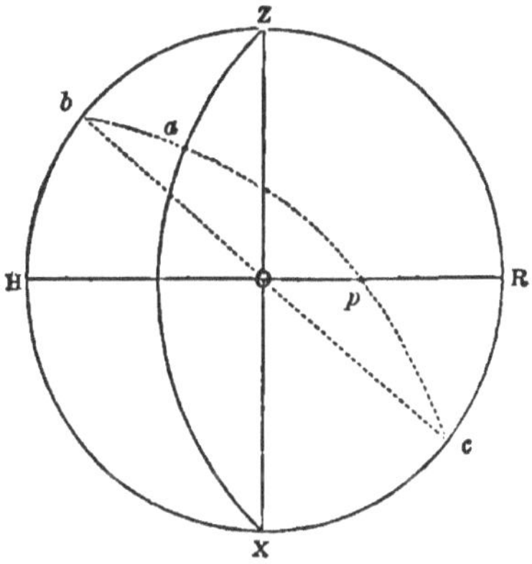

a great circle (Prob. IV.), and it shall be perpendicular to ZaX. If the perpendicular be required to pass through any point, as a, through the two points a and p, describe an oblique circle by Prob. IV.

56. PROB. XI.—*To draw a great circle which shall make an angle of, say 30°, with the primitive.*

Draw a diameter from Z at right angles to HR from the point Z as a centre, and with the chord of 60° from the plane scale by which the circle was drawn describe the arc OEG. From the centre O make OE (on OEG) equal to 30° on the scale of chords. Join Z E and produce to *x*, and *x* will be the centre of the oblique circle Z Y X, and it measures 30° on the scale of semi-tangents, *counting from* 90° on the

Fig. 35.

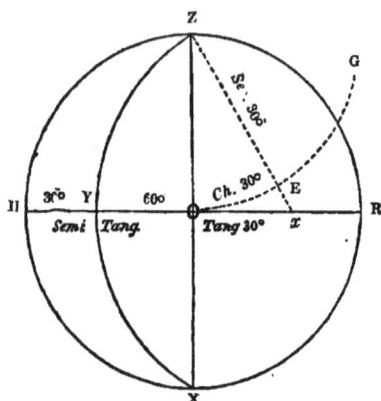

scale, or from H towards O. Thus any oblique circle may be drawn by taking the number of degrees from the scale of secants; for example, the radius of the oblique circle 30° (according to the problem) is the secant of 30°.

57. PROB. XII.—*To measure any part of an oblique circle (as a b in H a b R).*

Fig. 36.

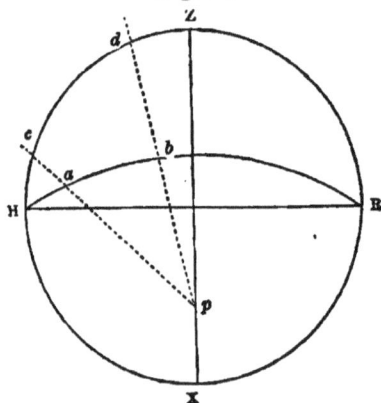

Find the pole of the oblique circle as at p (by Prob. VII.).
Join pa and produce to c, and join pb and produce to d,
and the measure cd on the scale of chords will be the
measure of ab.

Note.—By this problem any number of degrees may be
laid on an oblique circle.

58. PROB. XIII.—*To measure an angle at the primitive,
as HZa.*

N.B.—The angle at the primitive is always measured on
a right circle which lies 90° distant, or which passes through
the pole p of the oblique circle. Apply the distance Ha

Fig. 37.

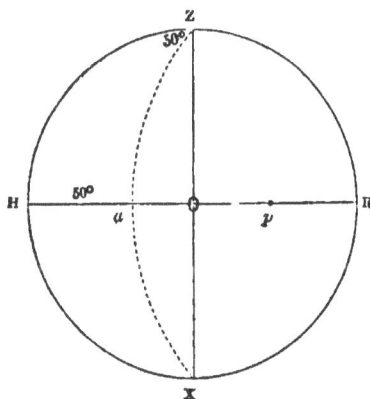

in the dividers to the scale of semi-tangents, counting backwards from 90°. Then Ha is the measure of HZa and =
50°.

N.B.—This problem is very useful to the navigator.

59. PROB. XIV.—*To measure the angle ZaR.*

Having found the pole of ZaX (by Prob. VII.) to be at
p, and the pole of HaR to be at p, join ap and produce

it to *b*, and join *ap'* and produce it to *c*, and the distance *bc* measured on the scale of chords (or 68°) will be the measure of the angle Z*a*R or H*a*X.

Fig. 38.

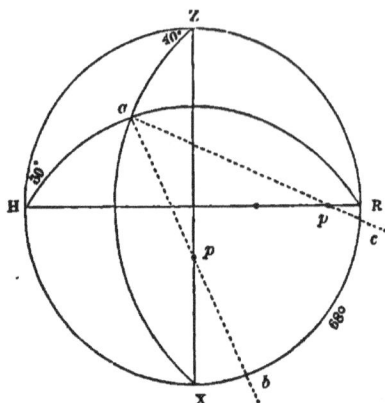

60. It must be remembered that the pole of a right circle is always at the primitive; thus the pole of ZX is at H or R (being at a distance of 90°), and the pole of HR is at Z or X.

N.B.—The right circle appears as a *diameter* in the stereographic projection.

CONSTRUCTION OF SPHERIC TRIANGLES.

61. In proceeding to the *construction* of spheric triangles the navigator must bear in mind that it is convenient to make arcs of latitude, declination, altitude, &c., to occupy the same relative positions in the sphere — that is to say, latitude is *always* on the primitive, &c., or in other words, he had better use a *right* sphere, as above, instead of an *oblique* sphere.

Declination is *always* a small circle parallel to the equator.

Altitude is *always* a small circle parallel to the horizon.

Indeed, the lines, as they are drawn on Fig. 23, the illustrative diagram, ought to be perfectly understood and remembered.

In all the figures following the *same letters* will designate the same parts. Thus—

HR will always represent the horizon.

PS the polar axis of the earth as prolonged or *produced* to the heavens (of which the primitive is the imaginary limit, *the earth itself being now supposed to be the very small point at the centre of each figure*).

ZN the zenith and nadir.

EQ the equator.

dc a parallel of declination.

ab a parallel of altitude.

x the position of the heavenly body.

While the answers to the problems will be indicated by a *thick line* at the part of the figure where the answer is to be measured.

Fig. 39.

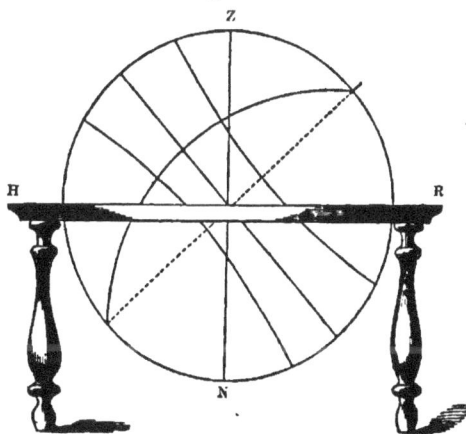

The line HR will consequently exactly coincide in posi-

E

tion with, and represent the wooden horizon of, the artificial globe (Fig. 39), while the other lines of the sphere correspond also with those on the globe, but seen as slightly distorted by the nature of the projection.

62. Those who possess a globe will do well to compare the following problems with the lines on it.

It is necessary to remember that a mere spheric triangle may be formed from any three parts given, and either a side or angle may be placed according to convenience in construction. But the limiting of certain data to certain parts of the projection is a mere conventional rule, in order to simplify the study to the minds of those who have neither time nor inclination to perfectly master the whole doctrine of spherics, but who desire a mere knowledge of its principles and practice as applicable to the wants of the navigator.

MEMORANDA.

63. All *azimuth circles* meet at the zenith and cut the horizon at right angles, and are measured along it.

All *hour circles* meet at the poles of the world, which are points on the primitive 90° from the equator.

A parallel of declination is a small circle which the sun or a heavenly body seems to describe round the pole.

The spheric figure in general use, although a hemisphere, really represents the whole sphere, inasmuch as the hour circles merely imply *time from noon*, A.M. or P.M. ; and, consequently, the hour circle for 10 A.M. answers for 2 P.M., each being two hours from noon. In like manner with azimuth circles, the point next to south may be either S. by E. or S. by W. according as we consider the centre as the east or west point.

Amplitude is distance of an azimuth circle from W. or E., as measured on the horizon.

Azimuth is distance of an azimuth circle from N. or S., as measured on the horizon.

Latitude is distance from the equator.

Longitude is distance from the meridian which passes through Greenwich Observatory.

64. As the numbers used in the following constructions are merely intended to serve the purpose of illustration, answers are given to the nearest degree only.

. The usually-occurring questions in nautical astronomy will first be answered by projection, and afterwards (107) the same figure will be repeated when working by calculation, such additions and arrangements being made to them as the process of calculation requires, in order to adapt them to it.

TO FIND THE LATITUDE OF A PLACE.

65. I.—Given, meridian altitude sun's centre, 60°.
      „ declination   .   .   . 20° N.
      (Observer north of the sun.)

Draw the circle with the chord of 60°.

Draw two diameters, H R and Z N, at right angles to

Fig. 40.

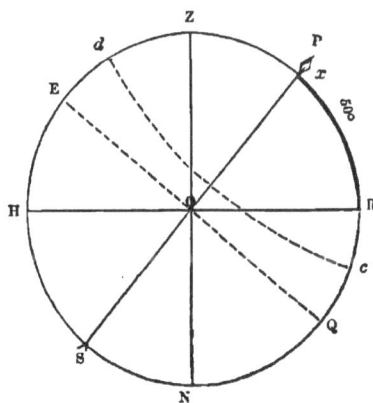

each other. (In the following questions the circle is supposed to be drawn and quartered.)

Lay off the given altitude 60° taken from the line of chords, from H on the horizon towards Z the zenith, say to d, and d will represent the sun's place on the meridian. The sun being in 20° N. declination the equator will be 20° southward of d, or at E. Join E O, and produce it to Q, and make the polar axis, P S, at right angles to it; then P R will be the height of the pole P, which is equal to the latitude 50°, as measured on the line of chords (45).

66. II.—Given, sun's altitude . 50°.

,,   declination   20° N.

,,   azimuth, S. 45° E.

Draw $a\,l$, the parallel of altitude 50° (parallel to H R), by Prob. IX. Draw the azimuth circle Z x N, 45°, from H,

Fig. 41.

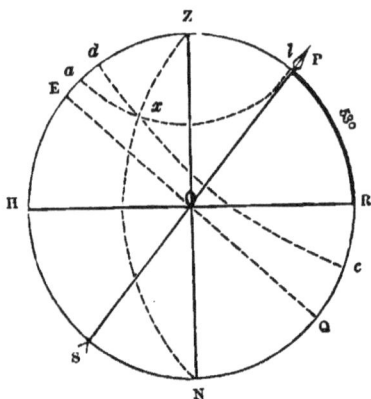

the south point of the horizon, by Prob. XI., making it 45° from the primitive: where these intersect will be the sun's place x.

Then with the secant of the complement of the declination, or 70°, intersect the tangent of 70° laid off in the

same direction northward of the equator (because declination is north), where these intersect will be the centre
from which with the tangent of 70° describe the small
circle $d\,c$ (by Prob. IX.); through this centre and the
centre of the primitive draw the polar axis, P S, and the
measure, R P, on the scale of chords will be the latitude.

67. III.—Given, declination, 10° S.
       ,,   altitude . 50°.
       ,,   time   . 10 A.M.

N.B.—In south latitude (being south of the sun) draw
the primitive with the chord of 60°.

In the primitive assume a point S (above what is intended to be the south point of the horizon), and quarter
the circle.

Fig. 42.

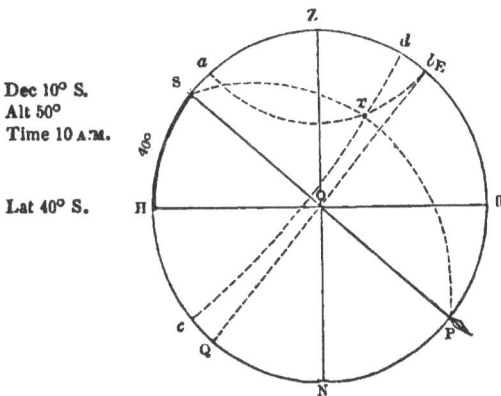

Dec 10° S.
Alt 50°
Time 10 A.M.

Lat 40° S.

Draw S x P, two hours or 30° from the primitive (Prob.
XI.).

Draw $d\,c$ = 10° on the south side of the equator by
Prob. IX. 10° from E Q.

Then with the secant from centre O, and tangent from

centre x of the co-altitude $= 40°$, find the centre of $a\,l$ (Prob. IX.), and draw it with tangent 40° as a radius; through the centre of this parallel and the centre of circle draw a diameter, Z O N, and another, H O R, at right angles to it, and the distance, H S, will be the latitude, 40° south.

68. IV. — Given, altitude of a celestial body on the meridian, below the pole, say,—

Altitude of a Lyræ . . 20° below the pole.
Declination of a Lyræ . 38½° N.

(Observer south of the star.)

Let x be the star's place, the declination being 38½°, the equator E Q, will be 38½° south of it, as measured on

Fig. 43.

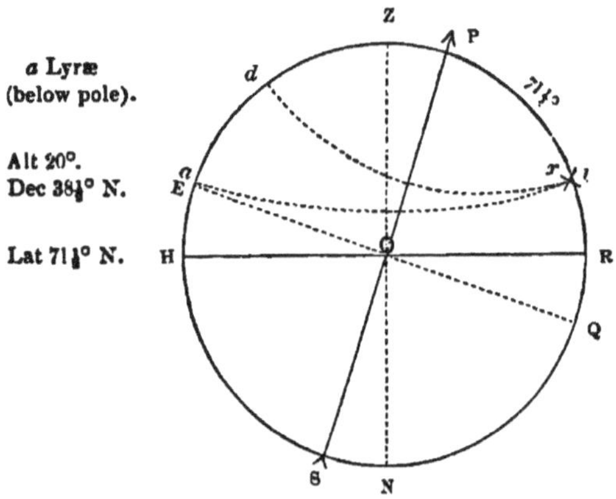

a Lyræ
(below pole).

Alt 20°.
Dec 38½° N.

Lat 71½° N.

the scale of chords, and the altitude, 20°, will give the horizon at R 20° below x. Draw H R and E Q, and diameters at right angles to each, and the height of the pole P above R will be the latitude, 71½°. The parallels of declination and altitude may be drawn by Prob. IX.

69. V.—Given, time   .   9 A.M.
     „   declination 20° N.
     „   azimuth S. 60° E.

(Observer north of the sun.)

Draw the circle with the chord of 60°.

Assume the polar point P, and draw the polar diameter, P O S, and also E Q, the equator, at right angles to it.

Draw the hour circle, 9 A.M. = 45° from the primitive (by Prob. XI.). Draw $d\,c$, the declination (by Prob. IX.), then the intersection, x, is the sun's place.

Fig. 44.

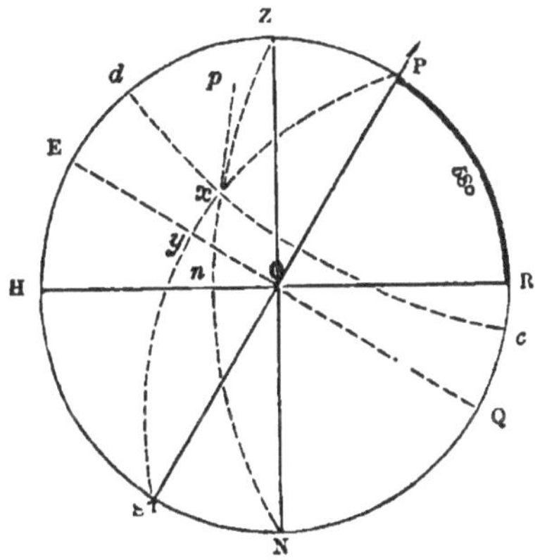

Time, 9 A.M.
Dec. 20° N.
Az. S. 60° E.

Lat 58° N.

The given azimuth circle is 60° from south, and the azimuth circle passing through x must be drawn by Prob. VIII., as $Z\,x\,N$; lay off 90° on the scale of chords, from Z to R, and R P will measure the height of the pole or be the latitude = 58° N.

TO FIND THE TIME.

70. VI.—Given, latitude   . 45° 40′ N.

      „   declination . 10° N.

      „   azimuth . S. 45° E.

Draw H R and Z N at right angles to each other.
From the scale of chords lay off 45° 40′ from R to P.
Draw P S and E Q.
Draw the azimuth circle, Z x N, by Prob. XI., and the
parallel of declination, 10° N., by Prob. IX., then the
point of intersection, x, will be the sun's place.

Fig. 45.

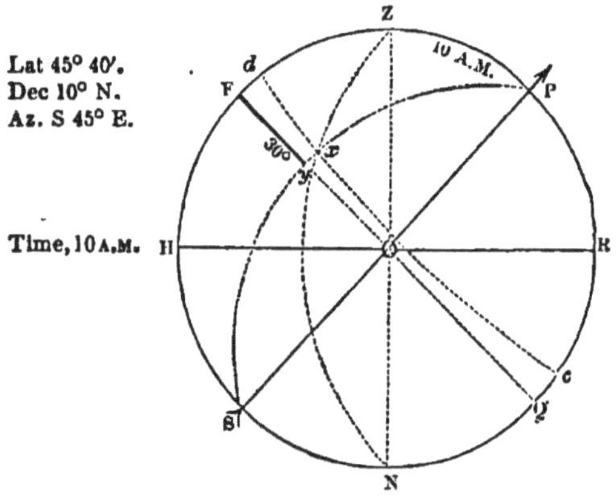

Lat 45° 40′.
Dec 10° N.
Az. S 45° E.

Time, 10 A.M.

Through P x S draw an oblique circle by Prob. IV.,
and the angle Z P x will be the hour angle, equal to two
hours from noon (or the meridian), or 30°, or 10 A.M.,
measured on E y, from semi-tangents, *backwards* from 90°.

71. VII.—Given, declination   . 20° S.

      „   altitude .   . 20°.

      „   azimuth . S. 45° W.

Draw H R and Z N.
Draw the azimuth circle by Prob. XI., as Z x N.

Draw the parallel of altitude, 20°, by Prob. IX., then *x* will be the sun's place.

Draw the parallel of declination, *d c*, through the point *x*, by Prob. IX.

By Prob. IV. draw P *x* S through the three points, and the angle Z P *x* will be the hour circle, and is measured on E *y*, a diameter at right angles to P S, and is equal to 45°, or three hours from noon, *westerly*, by the equator, or 3 P.M.

Fig. 46.

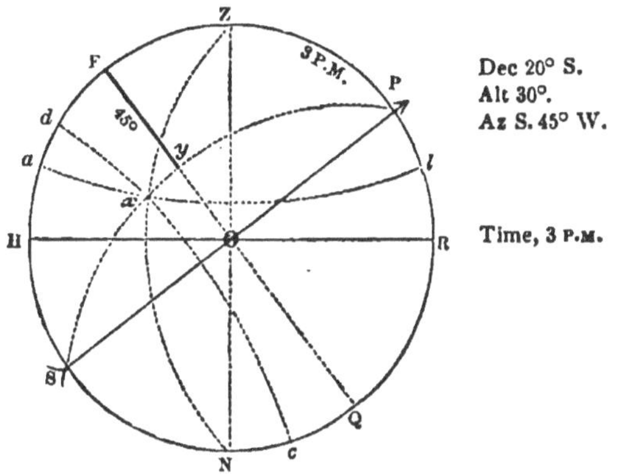
Z
F
3 P.M.
P
d
45°
y
a
l
x
H
O
R
S
Q
N
c
Dec 20° S.
Alt 30°.
Az S. 45° W.
Time, 3 P.M.

72. VIII.—Given, latitude . 21° N.
,, declination 20° S.
,, altitude . 30°.

Draw H R and Z N.

Make R P equal to 21° from the scale of chords.

Draw P S and E Q.

Draw the parallel of declination, *d c*, by Prob. IX.

Draw the parallel of altitude, *a l*, by Prob. IX., and the intersection *x* will be the sun's place; through the points P *x* and S draw the oblique circle (by Prob. IV.), and the angle Z P *x* will be the hour angle, and measured on

E y = 45°, or three hours from noon, being 9 A.M. or 3 P.M.

Fig. 47.

Lat 21° N.
Dec 20° S.
Alt 30°.

Time, 9 A.M.

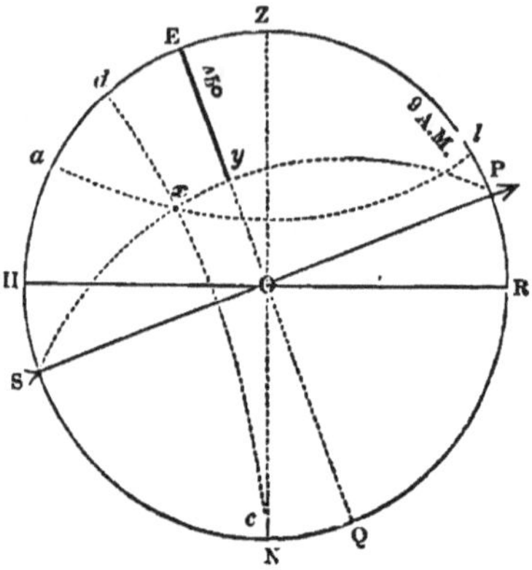

TO FIND AN AZIMUTH.

73. IX.—Given, latitude . 36° S.
„ declination 20° N.
„ altitude . 20°.

Draw H R and Z N.

Lay off the latitude from the scale of chords = 36° from the south point of the horizon at H to S.

Fig. 48.

Lat 36° S.
Dec 20° N.
Alt 20°.

Az N. 45° E.

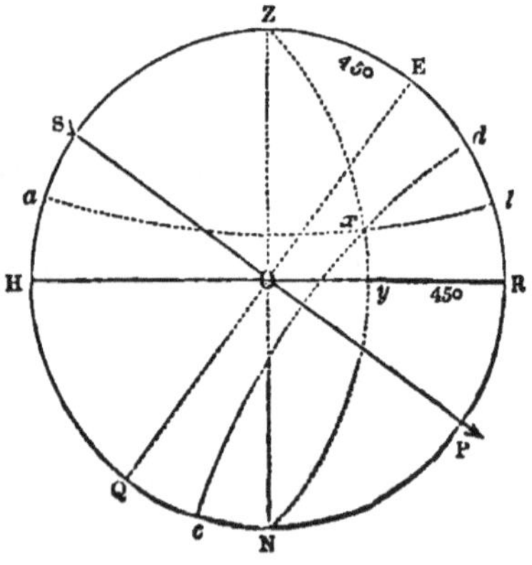

Draw S P and E Q.

Draw declination north by Prob. IX. (as dc), and

Draw $a\ l$, the parallel of altitude, 20° by Prob. IX., then x is the sun's place.

Through the three points, Z x and N (by Prob. IV), draw an oblique circle, Z y N, and the angle y Z R will be the azimuthal angle $= 45°$, as measured at R to y, on the semi-tangents *backwards* from 90° on the scale.

74. X.—Given, latitude . 21° N.

    „   time . . 9 A.M.

    „   declination 20° S.

Draw H R and Z N.

Make R P equal to the latitude 21°.

Draw P O S and E Q.

Fig. 49.

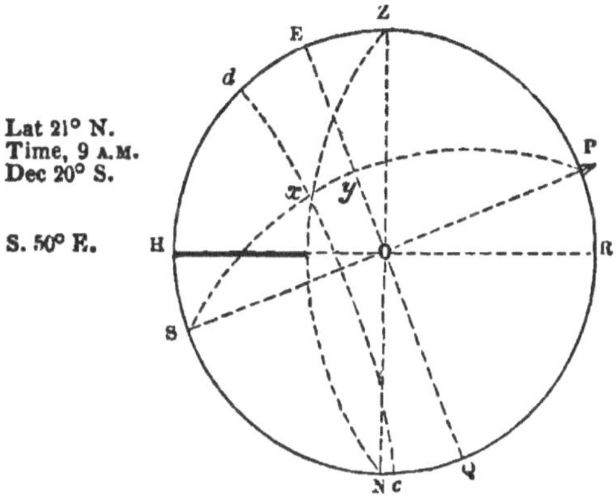

Lat 21° N.
Time, 9 A.M.
Dec 20° S.

S. 50° R.

Draw the hour circle, three hours or 45° from noon (or the primitive), by Prob. XI., as E y.

Draw the parallel of declination, $d\ c$, south (by Prob. IX.), the point of intersection, x will be the sun's place; through the points Z x N draw an oblique azimuth circle by Prob. IV., and the angle x Z R will be the azimuth from north, or N. 130° E., and the angle H Z y will be the azimuth from south, or S. 50° E.

CALCULATION OF SPHERIC TRIANGLES.

75. In proceeding to the calculation of spheric triangles, we notice that such are either right angled (*i.e.* having one angle equal to 90°), quadrantal (*i.e.* having one side equal to 90°), or oblique (*i.e.* having neither an angle nor a side equal to 90°).

1. RIGHT-ANGLED SPHERIC TRIANGLES.

76. Every triangle, as in plane triangles, has " six parts," viz., three sides and three angles; and any three of these being given, the rest may be found by proportion.   But in a right-angled spheric triangle two parts only need be given besides the right angle.

77. In calculating parts of a triangle, whether plane or spherical, *we shall often save much trouble if we consider, first, whether of the three things or parts given any two of them are a side and an opposite angle.*   When such is the case the " rule of sines," as it is called, founded on the fundamental theorem that " the sides of a triangle are in proportion to the sines of their opposite angles " is peculiarly simple.

To find a Side.

78. RULE.—As the sine of any given angle is to the sine of its opposite side, so is the sine of any other given angle to the sine of its opposite side.

To find an Angle.

79. RULE.—As any given side is to the sine of its opposite angle, so is any other given side to the sine of its opposite angle.

It is not, as already declared, the purpose of this book to do more than give a plain but substantially practical

introduction to the study of spherics, leaving the argu‑
mentative proofs of various theorems to the few works on
the subject which are already before the public, or to a
succeeding volume.

80. But that which above has been called a "funda‑
mental" rule deserves, in passing, a little attention, be it
only to encourage the student towards further research; as
in this he will see the simplicity of the study of geometry
when it is approached by a proper path.

In the following plane triangle A B C bisect each side
and erect perpendiculars; they will meet in O, the centre

Fig. 50.

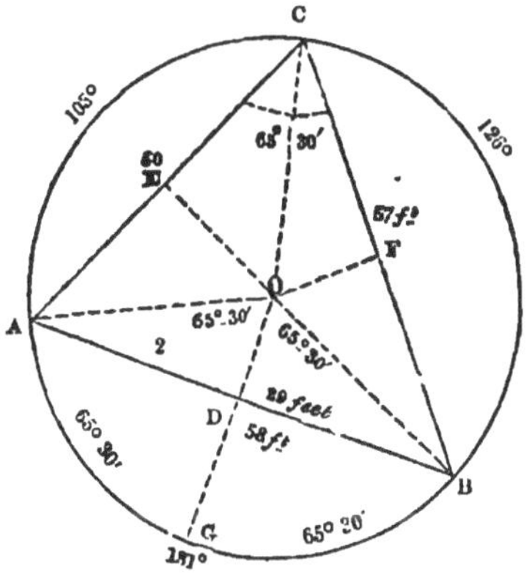

(N.B.—Lengths and angles are marked in order that the student may
easily verify by logarithms).

of the circumscribing circle. Join O A, O B, and O C. Now,
by reference to page 19, we shall find that Euclid, in Book
III., Prob. XX., proves that "*an angle at the centre of a
circle is double the angle at the circumference upon the
same base;*" therefore, in the above figure, the angle A O B
is double the angle A C B; but by construction A D is the
half of A B; similarly, the angle A O D is the half of the
angle A O B; therefore, the angle A O D equals the angle
A C B.   Now, A B, the base of angles A O B and A C B, is a
chord of the arc A C B, and A D being half of A B (being by

definition called a "sine"), subtends the angle A O D or A C B.

Hence we find A D equals the sine of the angle A O D—equals the sine of the angle A C B.

By taking another base as A C, and another as C B, we shall, by the same method of demonstration, find that E C is equal to the sine of angle C B A, and also that F B is equal to sine of the angle B A C, and putting a for the side B C, and b for the side A C, and c for the side A B, we shall have the following equations :—

$$\tfrac{1}{2}\, a = \text{sine A } (\textit{i.e.} \text{ sine of } \angle\text{A})$$
$$\tfrac{1}{2}\, b = \text{sine B}$$
$$\tfrac{1}{2}\, c = \text{sine C}$$

And by combination :—

$$\tfrac{1}{2}\, a : \tfrac{1}{2}\, b :: \text{sine A} : \text{sine B,}$$

Or, $a : b :: \text{sine A} : \text{sine B, &c.}$

Or, $a : \text{sine A} :: b : \text{sine B, &c.}$

Thus the sides are in proportion to the sines of these opposite angles.

A number of useful formulæ, which seem to wear so forbidding an aspect in ordinary works upon Trigonometry, are really nothing more than easily obtained deductions from the above.

THE FIVE CIRCULAR PARTS.

81. When, however, the calculation of right-angled spheric triangles cannot fall under the rule above given (from having no angle and opposite side given), a method invented by Lord Napier and published in 1614, and which is called the "Circular Parts" (or because the right angle is never considered one of them, is called also the "Five Circular Parts") is singularly applicable to all cases which can occur.

82. Anyone of these five circular parts may be considered the *middle* part, the parts joining thereto being called *extremes conjunct;* but the parts which are *separated* by an angle or a side are called *extremes disjunct.*

N.B. The right angle does not separate its two containing sides.

83. In every case, then, as three things must enter into every consideration of proportion, viz. the two given parts (excluding the right angle) and the part required, one must be called a *middle,* while the others are considered as *conjunct* or *disjunct,* as the case may be, but in using the part in computation it must be remembered that—

> " When angles or hypothenuse
> Among the parts you trace,
> Their complements
> Or supplements
> Must always take their place."

(A quaint rhyme or two may aid the memory.)

84. The manner in which the equations are formed from which we derive the proportions may be thus explained :—

Fig. 51.

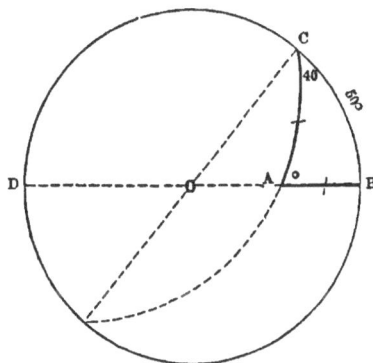

Draw any spheric triangle A B C.

Then, because the right circle A B passes through O the

pole of the primitive of which one side B C of the triangle
is an arc, the angle B is a right angle.

The " Five Parts " are, therefore (83) :—

    1. The complement of the hypothenuse A C.
    2. The side A B
    3. The side C B
    4. The complement of angle A
    5. The complement of angle C.

Now, suppose, in the above triangle, the sides A C and
A B are given to find the ∠ A.

The middle part must be so selected as to make the
other parts either disjunct or conjunct (not one conjunct
and the other disjunct). The middle part in this triangle
will evidently be the ∠ A, as the two given sides include
it, and because they *join it* they will be extremes conjunct.

N.B. The hypothenuse is always the side opposite the
right angle. Calculation then depends on the following
universal, or as it has been long called the " catholic proposition," viz. :—

85. The sine of the middle part multiplied into radius is reciprocally
proportional with the tangents of extremes conjunct, and with the cosines
of extremes disjunct.

Expressed as an equation it would stand thus:—

sine of middle × radius = tan extr conjunct × other extr conjunct.

Remembering that of four numbers in proportion, the
product of the means always equals the product of·the extremes (see page 21), we may vary the above as follows,
viz. :—

radius : tan extr conjunct : : tan other extr conjunct : sine of middle.
or, tan extr conjunct : radius : : side of middle : tan other extr conj.

If the extremes are disjunct we have as follows :—

sine of middle × radius = cos extr disjunct × cos other extr disjunct.

or, radius : cos extr disjunct : : cos other extr disjunct : sine of middle ;
or, cos extr disjunct : radius : : sine of middle : cos other extr disjunct.

It follows, then, that as we can only want, in any case of right-angled spheric trigonometry, to find either a middle part or an extreme, the following four simple formulæ are all-sufficient :—

If extremes are conjunct.

RULE A. Sine of middle = $\dfrac{\text{tang extr conjunct} \times \text{tang other extr conjunct}}{\text{radius.}}$

RULE B. Tang extr conjunct = $\dfrac{\text{radius} \times \text{sine middle part}}{\text{tang other extr. conjunct.}}$

If extremes are disjunct.

RULE C. Sine of middle = $\dfrac{\text{cos extr disjunct} \times \text{cos other extr disjunct}}{\text{radius.}}$

RULE D. Cos. extr disjunct = $\dfrac{\text{radius} \times \text{middle part}}{\text{cos other extr disjunct.}}$

Or, as adapted at once to logarithmic calculation, we can put the above still more plainly, thus :—

When extremes are conjunct.

86. RULE a. To find log sine middle $\left\{ \begin{array}{l} \text{From log tang extr conjunct + log tang} \\ \text{other extr conjunct subtr rad (or 10)} \end{array} \right.$

87. RULE b. To find log tang extr conjunct $\left\{ \begin{array}{l} \text{From log rad + log sine of middle} \\ \text{subtr log tang other extr conjunct} \end{array} \right.$

When extremes are disjunct.

88. RULE c. To find log sine middle $\left\{ \begin{array}{l} \text{From log cos extr disjunct + log cos} \\ \text{other extr disjunct subtr rad} \end{array} \right.$

89. RULE d. To find log cos extr disjunct $\left\{ \begin{array}{l} \text{From rad + log sine middle subtr} \\ \text{log cos other extr disjunct} \end{array} \right.$

care being taken to use complements of angles and hypothenuse. We will take an example, and at once apply the above formulæ.

90. In the right-angled triangle suppose the following:—

Given, a side A C = 30°
       a side B C = 40 } required, hypothenuse A B and ∠ A.

Fig. 52.

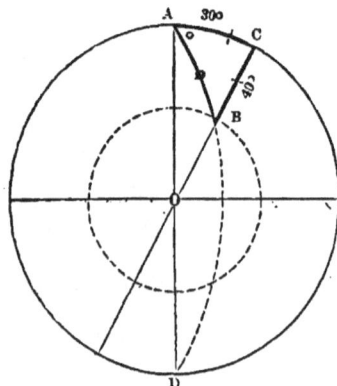

In constructing a spheric figure, it is, moreover, always convenient to place a given side on the primitive: we do so by making A C = 30° (by scale of chords) join C O.

By Problem V. draw a small circle parallel to the primitive = 40° distance; and where this cuts O C will be the third angular point B. Through the points A B D draw a great circle (by Prob. IV.), and A B C will be the triangle, of which the ∠ C will be a right angle (because the right circle C O passes through O, the pole of the primitive, and therefore is perpendicular to the arc A C), and its opposite side will be A B the hypothenuse.

N.B. It is convenient to mark the given parts as in the figure, and the parts required with an *o*.

To find the hypothenuse A B.

It will be seen that in order to get two "parts" *alike* (that is, two parts which shall be either *conjunct* or *disjunct*), the middle part should in the above figure be the hypothenuse itself, for it is separated from the parts given by

∠ A at one end, and by ∠ B at the other. The parts
A C and B C are therefore called extremes *disjunct.*

91. The following rhymes may help the memory in
applying tangents or cosines, &c. : —

> Tangents join the middle
> (Put the middle where you please) ;
> Cosines afar,   ·
> From middle are
> (Five parts you have in these).

We have, therefore, in this example, to use *cosines* with
the extremes (subject to the correction for hypothenuse and
angles), and we want to find the middle part, A B. Rule
C gives the following equation :—

$$\text{Sine of middle part} = \frac{\cos \text{ ext. disj. } \times \text{ cos of other ext. disj.}}{\text{radius.}}$$

Now, before proceeding, let us consider what is meant
by this equation, and why it was further altered into rule *c*
(88). We have already shown that multiplication is performed in logarithms by addition ; and division by subtraction ; and in the fraction standing on the right side of the
sign of equation, there are two quantities to be multiplied,
and a quantity which is to divide their product. We therefore *add* the logarithms of the two factors in the numerator from the sum and *subtract* the logarithm of the
denominator.

92. In a proportion worked by logarithms it is better to
place the terms vertically (putting the divisor as the first
term), thus :—

> As radius
> is to cos extreme disjunct,
> so is cos of the other extreme disjunct
> to the sine of the middle part.

(Remember old Dr. Kelly's Hibernian rhyme : —)

> "Now the product of radius and middle part sine,
> Equals that of the tangents of parts that combine,
> And also the cosines of those that dis*join*.")

93. But we have first (83) to correct the above proportion if the hypothenuse or an angle form part of it.   It should therefore appear accurately thus :—

as radius   .   .   .   co ar 10·000000
is to cos side A C 30°   .   .   9·937531
so is cos side B C 40°   .   .   9·884254
                                        ——————
to cos hyp. A B   .   .   . =9·821785 = 48° 26′ 21″

To find ∠ A :—Use the rule of sines (79), (having now opposite sides and angles).

as sine A|B 48° 26′ 21″   .   .   co ar   0·125952
to sine of opp. ∠ (radius)   .   .   10·
so is sine of B C 40°   .   .   .   9·808067
                                            ——————
to sine of ∠ A   .   .   .   .   9·934019 = 59° 12′ 37″

But as it is better to thoroughly master one question in all its bearings, we will take a different view of the same question, and determine on finding the angle A, before we find the hypothenuse.

We now evidently call A C the middle, and then the ∠ A and the side B C will be extremes *conjunct* (the right angle is not one of, and *does not separate* the parts remember), and Rule 87 gives us as an equation :—

To find log tang extr conjunct { From log rad + log sine of middle subtr
                                     log tang of other extr conjunct

tang of the other ext. conj. B C 40°   .   co. ar. 0·076186
rad   .   .   .   .   10·
sine of middle A C 30°   ↖   .   .   .   9·698970
                                                     ——————
co tang ∠ A   .   .   .   .   .   .   9·775156 = 59° 12′ 37″

The hypothenuse can now be found by the rule of sines (78), thus :—

as sine ∠ A 59° 12′ 37″   .   .   co ar   0·065981
to sine side B C 40° .   .   .   .   9·808067
sine 90°   .   .   .   .   .   .   10·
                                                  ——————
sine hyp.   .   .   .   .   .   9·874048 = 48° 26′ 21″

94. The co. ar. (read arithmetical complement) of an arc is what the logarithm wants of radius, and is readily formed by subtracting each figure of the logarithm (*beginning at the left hand*) from 9 and the last from 10 : thus the logarithmic co. ar. of ·333333 is ·666667. This saves subtraction as the three logs may then be added.

Every right-angled spheric triangle may in like manner be worked by the four equations A, B, C, D, or *a, b, c, d* (page 65). But in such triangles as have a side for a right angle, a modification of the above rules is necessary, for we have in such cases what are called

QUADRANTAL SPHERIC TRIANGLES.

95. The only difference in the mode of working arises from an apparently whimsical perversion of terms.

> For now the merry *quadrant*
> Its pranks with us to play,
> *Transforms itself to radius,*
> And laughs our rules away.
> It calls legs, angles !—angles, legs !
> Our notions to confuse ;
> While its opposite angle's supplement,
> It calls hypothenuse !

So that, considering the quadrantal side as radius, and the supplement of its opposite angle as hypothenuse, the solution of quadrantal angles is performed by rules already explained ; viz., the rule of sines, and the four rules for the five circular parts.

OBLIQUE SPHERIC TRIGONOMETRY.

96. The majority of spheric questions which occur in practice fall under the denomination "oblique," *i. e.* having neither an angle nor a side equal to a right angle.

97. Oblique spheric trigonometry admits of the six following cases, viz. :—

The given parts will be either

    1. Two sides and an opposite angle.
    2. Two angles and an opposite side.
    3. Two sides and an included angle.
    4. Two angles and an included side.
    5. Three sides.
    6. Three angles.

98. The solution of the first four cases may either be effected by means of a perpendicular let fall from one of the angles to its opposite side, or by special rules not requiring the perpendicular. There is a little difficulty with beginners in constructing the triangle so as to admit of a perpendicular being drawn in *such manner as to retain two of the given parts in one of two new right-angled triangles thus formed.* This may often be avoided by attending to the following directions, viz., describe the usual circle and quarter it. Then lay off a given side, A C, on the primitive · (A being the angular point of the *left* hand of the side of the triangle, which lies on the primitive, and C the other end of it. It is merely convenient to have one method), and at C lay off the given angle (Prob. XI). Consider how you can secure two of the given parts in a new right angle you are about to form, and (by Prob. X.) let fall the perpendicular accordingly.

99. When the two angles which lie upon the side which is to be crossed by a perpendicular are of *like affection,* i. e., both greater or both less than a right angle, the perpendicular will fall *within* the triangle, but when unlike, *i. e.* when one is *acute* and one *obtuse,* the perpendicular will fall *without* the triangle.

100. N.B. In explaining the nature of " construction " from observation, it was recommended that the various circles of the sphere should always be made to represent their respective parts in the general astronomic diagram, but for calculation it is better to take the parts merely as sides or angles.

The following will show the method of "projecting" and calculating any oblique spheric triangle which can possibly occur.

101. CASE I.

Two sides and an opposite angle.

Given, a side 60°
    ,,   a side 100°   } To find a side and the other angles.
    ,,   an opposite angle 130°

By construction :—

N.B. In nearly all cases we suppose a circle to be already drawn with a chord of 60°, and two diameters at right angles within it.

Fig. 53.

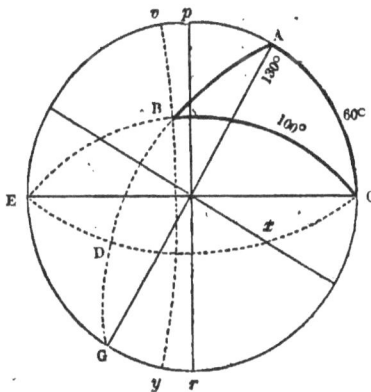

Lay off from C to A the one side, 60° (from scale of chords). From A draw A G, with the supplement of 130°, and draw (by Prob. VIII.) the great circle A B G. About E describe the small circle $v\,y$, at a distance of the supplement of 100° (or 80°) from it (by Prob. VI.); and where $v\,y$ cuts A B G is the angular point B.

With the three points E B C draw an oblique circle (by Prob. IV.), and A B C is the triangle.

To **measure** the parts required, viz., A B, ∠ C, and ∠ B:

       A B is measured by Prob. XII.
       ∠ C is measured by Prob. XIII.
       ∠ B is measured by Prob. XIV.

By calculation :—

     1st. By means of a Perpendicular.

Having a side on the primitive with an adjacent ∠ A given, let fall a perpendicular from the ∠ C upon AB (produced if necessary) to D (Prob. X.).

To find the other opposite ∠ ABC by rule of sines (79).

as sine BC 100° .	. . co ar	0·006649
to sine ∠ A 130°	. .	9·884254
so is sine A C 60°	. .	9·937531
to sine ∠ ABC	. .	9·828434 = 42° 21′

$$\begin{array}{r} 180 \\ \hline \angle\ CBD = 137\ \ 39 \end{array}$$

To find A D (in triangle A D C):—

The ∠ A will be the middle part, and the hypothenuse A C and the side A D will be extremes conjunct (87).

as cotang hyp A C 60°	co ar	0·238561
to rad	.	10·
cos middle ∠ A 130°	.	9·808067
to tang	.	10·046628 = 48° 4′

$$\begin{array}{r} 180 \\ \hline \text{tang A D} = 131\ \ 56 \end{array}$$

To find D B (in triangle C B D) :—

The ∠ B will be middle, and hyp B C and the side D B will be extremes conjunct (87).

```
as tang BC 100°     co ar 0·753681
to radius  .    .   . 10·
cos ∠ B 137° 39'    . 9·868670
tang DB  .    .     . 10·622351 = 76° 35'
```

To find ∠ C by rule of sines (79).

```
AD = 131° 56'     as sine AC 60°  .   . 0·062469
DB =  76   37     to sine ∠ B 42° 21'  . 9·828434
BA =  55   19     so is sine BA 55 19  . 9·915035
                  to sine ∠ C  .   .   . 9·805938 = 39° 46'
```

2nd. Without a Perpendicular.

When the solving of a spheric triangle presents any difficulties to the unpractised as to where to let fall the perpendicular, if time is precious, recourse may be had to the following rule :—

When two sides and an opposite angle are given.

First find the angle opposite to the other of the two given sides, and the third angle of the triangle may then be found thus :—

RULE.—As the sine of half the difference of the two sides
is to the sine of half their sum
so is the tangent of half the difference of the two angles
to the cotangent of half the contained angle.

Or, as applied to the preceding question, viz. :—

```
Given, a side AC =  60°
       a side BC = 100°
       ∠ A       = 130°
       ∠ B       =  42° 21'
```

To find the angle C :—

side A C	60°	as sine half diff 20° .	. co ar	0·465948
side B C	100	to sine half sum 80° .	. .	9·993351
	160	tang half diff angles 43° 49½′ .		9·982182
half sum	80	cotang half contd ∠ C 19° 53′	=	10·441481
half diff	20	∴ whole ∠ C is 39° 46′ .	.	

∠ A 130°

∠ B	42	21
	87	39
half diff	43	49½

Then the side AB may be found by the rule of sines.

102. CASE II.

Two angles and an opposite side.

Given, an angle 60° ⎫ To find a side
angle   70° ⎬   a side
opp. side 50° ⎭   an angle

By construction :—

At the point C make an angle of 60° with the primitive (Prob. VIII.) by drawing C B E ; about C, at the distance of

Fig. 54.

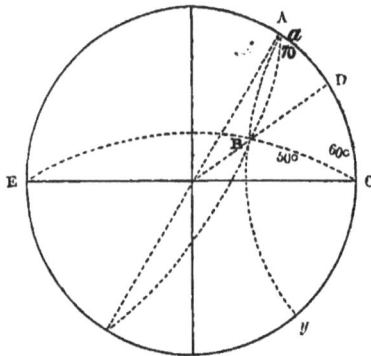

50° (the given side), draw the small circle A y (by Prob. IX.); at B, where these intersect, draw (by Prob. XI.) the great circle A B, making an angle of 70° with the primitive, and A B C will be the spheric triangle.

To **measure** the parts required, viz. A C, A B, and $\angle$ B :—

1. measure A C on the scale of chords   = 49° 37'
2. A B is measured by Prob. XII.   . = 44 55
3. $\angle$ B is measured by Prob. XIV.   . = 110 52

By calculation :—

1st. By means of a Perpendicular.

In order to preserve the two parts BC and $\angle$ C in the same rectangle, let fall a perpendicular on side A C. This is done by drawing a line from the centre of the circle, which is always the pole of the primitive, through the $\angle$ B till it cuts A C at D. (Prob. X.)

To find the other opposite side A B by rule of sines :—

  as sine $\angle$ A 70°   . co ar 0·027014
  to sine side BC 50° .   . 9·884254
  so is sine $\angle$ C 60   .   . 9·937531
  to sine side A B   .   . $\overline{9·848799}$ = 44° 54' 35''

To find the segment CD in $\triangle$ B C D :—

The angle C will be " middle," and the hypothenuse B C and C D will be extremes conjunct (87).

  as cotang hyp B C 50°   . co ar 0·076186
  to rad   .   .   .   .   10·
  so is cosine $\angle$ C 60°   .   . 9·698970
  to tang side C D   .   . $\overline{9·775156}$ = 30° 47' 23''

To find segment A D in triangle A B D :—

The angle will be *middle*, and the side A D and hyp. A B will be extremes conjunct (87).

  as cotang A B 44° 54' 35''   co ar 9·998631
  to radius   .   .   .   10·
  so is cos $\angle$ A 70°   .   . 9·534052
  to tang segment A D   .   . $\overline{9·532683}$ = 18° 49' 36''
  seg C D = 30° 47' 23''
  seg A D = 18 49 36
      $\overline{}$
      49 36 59 = side A C

To find ∠ B in △ABC by rule of sines (79).

as sine side BC   = 50° .   co ar 0·116746
to sine ∠ A   = 70   .   . 9·972986
so is sine side AC = 4937 . 9·881799
to sine ∠ B   9·970531 = 69°   7'   51"
                                      180
                          ∴ ∠B = 110   52   9

N.B. We take the supplement of 69° 7' 51" because the construction
shows the ∠ B to be obtuse.

2nd. Without a Perpendicular.

The side AB opposite the other given angle can be found
by rule of sines as before, and equals 44° 54' 35". Then
find ∠ B by the special rule given in Case I. thus:—

a side AB   44° 54' 35"   ∠ A   70°
a side BC   50   ∠ C   60
            ─────────────   ─────
            2)94 54 35   2)10
half sum of sides 47 27 18   half diff two angles   5
diff two sides   5   5 25
half diff two sides 2 32 42

as sine ½ diff 2 sides   2° 32' 42"   co ar 1·352577
to sine ½ their sum   47 27 18   .   9·867318
so is tang ½ diff 2 angles 5   0   0 .   8·941952
to cotang ½ contained ∠ B   10·161847 = 34 33 46
                                                        2
                                              ──────────
                                              69   7 32
                                              180
                                     ∠B = 110 52 28

Find A C by rule of sines (78).

as sine ∠A 70°   .   .   . co ar 0·027014   .
to sine side BC 50°   .   .   . 9·884254
so is sine ∠ B 110° 52' 28"   .   . 9·970516
to sine side A C   .   .   . = 9·881784 = 49° 36' 51"

103. Case III.

Two sides and an included angle.

Given, a side A C = 60°   ⎧ To find side B C.
  „   a side A B = 110°   ⎨   „   ∠ B
  „   the included angle A 45° ⎩   „   ∠ C

By construction :—

Lay off A C = 60° from the scale of chords.

Draw the great circle A D B (by Prob. XI.), at the distance of 45° from the primitive; about G draw the parallel circle $v\,y$, distance of the supplement of 110 (Prob. V.); where the two circles cross is the point B, through the points C B H draw a great circle (Prob. IV:), and A B C will be the triangle.

To **measure** the parts required :—

1. The ∠ C, $a\,x$ is measured on the scale of semi-tangents (Prob. XIII.).
2. measure ∠ B by Prob. XIV.
3. measure side C B by Prob. XII.

Fig. 55.

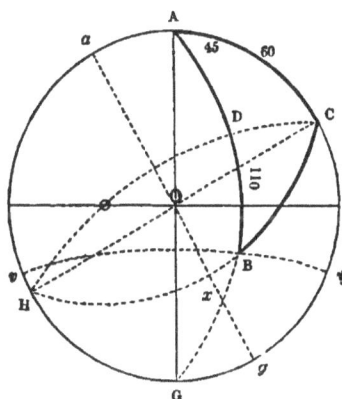

By calculation :—

1. *By means of a Perpendicular.*

Draw a perpendicular from the ∠ C upon the side A B to D (by Prob. X.).

Find the segment A D in the △ A C D (78).

Then the ∠ A is the middle, and hyp A C and side
A D are extremes conjunct (87).

```
as cotang hyp A C 60°  .   .   .   . co ar  10·238561
to radius        .   .   .   .   .   .   .  10·
so is cos ∠ A 45°    .   .   .   .   .      9·849485
                                          ──────────
  to tang of segment A D .   .   50° 46′ 17″ = 10·088046
  subtr from side AB      .   .  110
                                 ──────────
  segment .   .   .   .   .       59  13  43  = segment AD
```

Find D C in △ A D C by rule of sines (78):—

```
as rad.   .   .   .   . co ar  10·000000
sine side (hyp.) A C 60°  .   .  9·937531
sine ∠ A 45°  .   .   .   .      9·849485
                                ──────────
  to sine side D C  .   .   .    9·787016 = 37° 46′
```

Find ∠ B in △ B D C.
Here B D is middle, and D C and ∠ B are extremes
conjunct.

```
as tang D C 37° 46′     . co ar  0·110839
to rad.   .   .   .   .   .  10·
so is sine side B D 59° 14′  .   9·934123
                                ──────────
  to cotang ∠ B 42° 2′   .   .   10·044962
```

Find side B C by rule of sines (78):—

```
as sine ∠ B 42° 2′     . co ar  0·174209
to sine A C 60°    .   .   .     9·937531
so is sine ∠ A 45°  .   .   .    9·849485
                                ──────────
  to sine side B C   .   .   .   9·961225 = 66° 9′
```

Find ∠ C by rule of sines (79):—

```
as sine side A C 60° .   .   .   0·062469
to sine ∠ B 42° 2′ .   .   .     9·825791
so is sine side A B 110°  .   .  9·972986
                                ──────────
  to sine ∠ C  46° 36′  .   . = 9·861246
            180
         ──────────
  ∠ C = 133° 24′    N.B. We take the supplement because
                    the ∠ C is obtuse (by construction).
```

2. *Without a Perpendicular.*

RULE.—**When two sides and an included angle are given :—**

1. As the sine of half the sum of the two sides
is to the sine of half their difference
so is the cotangent of half their contained angle
to the tangent of half the difference of the other angles ;

and again,

2. As the cosine of half the sum of the two sides
is to the cosine of half their difference,
so is the cotangent of half the contained angle
to the tangent of half the sum of the other two angles.

And half the difference thus found added to half the sum gives the greater angle ; and half the difference subtracted from the half sum gives the smaller angle.

```
side A C =   60°
side A B = 110
          2)170
half sum    85        ∠ A (contained angle) = 45°
          2)50
half diff   25
```

Then by the above rules :—

as sine ½ sine of 2 sides 85°   .   . co ar   0·001656
to sine ½ diff ditto 25°   .   .   .   .   9·625948
so is cotang ½ contained ∠ A 22° 30'   .   10·382776
to tang ½ diff of other angles   .   .   . 10·010380 =   45° 41'
as cos ½ sum 85°   ,   ,   .   . co ar   1·059704
to cos. ½ diff. 25°   .   .   .   .   .   9·957276
so is cotang of ½ the cont. ∠ A 22° 30'   . 10·382776
to tang ½ sum of other angles   .   .   . 11·399756 =   87° 43'
                                 greater ∠ C = 133 24
                                 less ∠ B   =   42   2

Find B C by rule of sines (78).

as sine ∠ C 133° 24′ .   .   .   .   co ar   0·138720
to sine side A B 110° .   .   .   .   .   9·972986
so is sine ∠ A 45°   .   .   .   .   .   9·849485
                                        _____
                                        9·961191 = 66° 8′

104. Case IV.

Two angles and an included side.

Given, an angle 60°   ⎫ To find an angle
  „   an angle 36°   ⎬   a side
  „   included side 70° ⎭   a side.

Fig. 56.

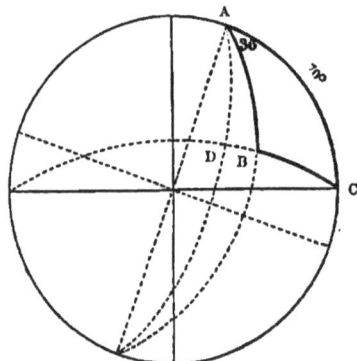

By construction :—

Lay off A C equal to 70° from the scale of chords.

Draw the great circle A B (Prob. XI.), at a distance of 36° .from the primitive, and in like manner C B at a distance of 60°; where these two intersect will be the point B, and A B C will be the triangle.

To **measure** the parts required :—

1. measure side A B by Prob. XII.
2. measure side B C   „   XII.
3. measure ∠ B   „   XIV.

By calculation :—

1. *By means of a Perpendicular.*

Let fall a perpendicular from ∠ A on side B C produced (Prob. X.) (or from ∠ C on A B produced, suppose the former).

Then will A D C be a right angle, as will also A D B right angled at D.

Find the ∠ D A C in the △ D A C.

The side A C will be middle, and the angle C, and ∠ D A C will be extremes conjunct (82): then by (87):—

as cotang ∠ C 60°	.	.	. co ar	0·238561
to rad	. . .	. .	.	10·
so is cosine side A C 70°	.	. .	.	9·534052
to cotang ∠ D A C	.	. .	.	9·772613 = 59° 21′
				∠ B A C − 36
				∠ D A B = 23 21

Find A D in △ A C D.

The angle D A C is middle, and extremes are conjunct.

as cotang side A C = 70°	.	. co ar	0·438934
to rad	. . .	. .	10·
so is cos ∠ D A C 59° 21′	.	. .	9·707393
to tang A D	. .	. .	10·146327 = 54° 28′

Find A B in △ A D B.

The ∠ D A B is middle, and sides A D and A B are extremes conjunct.

*as tang side A D 54° 28′	.	. co ar	9·853673
to rad	. . .	. .	10·
so is cos ∠ D A B 23° 21′	.	. .	9·962890
to tang side A B	. .	. .	9·816563 = 56° 45′

---

* It promotes accuracy to use the logarithm itself, from which the tang A D was taken on the previous calculation. 9·853673 is the ar. co. of 10·146327.

G

Find BC by rule of sines (78).

as sine ∠ C 60° .   .   co ar 0·062469
to sine side AB 56° 45'   .   9·922355
so is sine ∠ A 36°   .   .   9·769219
to sine side BC .   .   .   9·754043 = 34° 35'

Find ∠ ABC by rule of sines (79).

as sine side AB 56° 45'   co ar 0·077645
to sine ∠ C 60° .   .   .   9·937531
so is sine side AC 70°   .   9·972986
to sine ∠ABC .   .   .   = 9·988162 = 76° 41'
                                       180   0
                              ∠ B   103   19

N.B.—The supplement is used because construction shows the angle to be obtuse.

2. Without a Perpendicular.

RULE. — **When two angles and the included side are given :—**

As the sine of half the sum of the two angles
is to the sum of half the difference
so is the tangent of half the contained side
to the tangent of half the difference of the other two sides.

And again :—

As the cosine of half the sum of the two angles
is to the cosine of half their difference,
so is the tangent of half the included side
to the tangent of half the sum of the other sides;

and half the difference added to half the sum will give the greater side, and half the difference subtracted from half the sum will give the smaller side.

an angle   36°   side $\frac{70°}{2}$ = 35° = half the included side.
an angle   60
            2)96
half sum   48
            2)24
half diff   12

Find sides A B and B C.

as sine ½ sum 2 angles 48°	.	co ar 0·128927
to sine ½ diff 12°	. . .	9·317879
so is tang ½ included side 35°	. .	9·845227
to tang ½ diff other sides .	. .	9·292033 = 11° 5'
as cos ½ sum of 2 angles 48°	.	co ar 0·174489
to cos ½ diff 12°	. . .	9·990404
sine tang ½ included side 35°	. .	9·845227
to tang ½ sum other sides	. .	10·010120 = 45° 40'
		greater side A B = 56 45
		smaller side B C = 34 35

Find ∠ B by rule of sines (79).

as sine side B C 34° 35' .	co ar 0·245954
to sine ∠ 36° .	. . 9·769219
so is sine side A C 70°	. . 9·972986
to sine ∠ B	. . . 9·988159 = 76° 41'
	180
	∠ B 130 19

N.B.—The construction of the figure shows the ∠ B to be obtuse; therefore we use the supplement.

105. CASE V.

Three sides given : —

side A C   60°
side A B   70°
side B C   100°

Fig. 57.

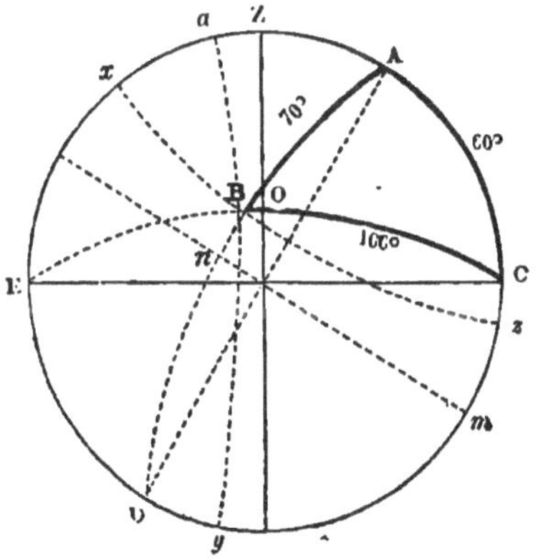

By construction : —

Lay off one side (say 60°) on the primitive, as at A C, from chord of 60°. About E draw (Prob. VI.) a small circle ay at the distance of the supplement of B C. About A draw a small circle xz at the distance of 70° (Prob. VI.). Through point B, where the small circles intersect, draw (by Prob. IV.) an oblique circle A B, and A B C will be the triangle.

To **measure** the parts required :—

1. mn on the scale of semi-tangents measures ∠ A.
2. The ∠ C is measured from Z to o (semi-tangent backwards from 90°).
3. The ∠ B by Prob. XIV.

By calculation : —

RULE.—Find half the sum of the three sides. Subtract from this half sum each of the two sides which, together, contain the required angle. Then add the sines of these two remainders to the sines of the two sides which contain the angle (using the co arcs of the latter). Half the sum of these four logarithms will give the sine of half the required angle.

To find the angle A :—

side A C	=	60°	sine	co ar	0·062469
side A B	=	70°	sine	co ar	0·027014
side B C	=	100°			

2)230

half sum of sides	115			
side A C	=	60		
		55	first remainder	9·913365
		115		
side A B	=	70		
		45	second remainder	9·849485

2)19·852333

sine 57° 32′ = 9·926166

2

∠A = 115 4

Find ∠ B by rule of sines (79).

as sine side B C 100°	.	.	co ar	0·006649
to sine ∠ A 115° 4'	.	.	,	9·957040
so is sine side A C 60°	,	,	,	9·937531
to sine ∠ B 52° 48'	.	.	.	=9·901220

Find ∠ C by rule of sines (79).

as sine side B C 100°	.	.	co ar	0·006649
to sine ∠ A 115° 4'	.	.	.	9·957040
so is sine side A B 70°	.	.	.	9·972986
to sine ∠ C 59° 48'	.	.	.	=9·936675

106. CASE VI.

Three angles given :—

Angle A 130°
Angle B 50°
Angle C 45°

Fig. 58.

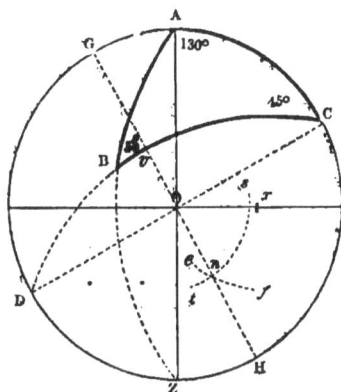

By construction : —

Make the ∠ A = 130° (by Prob. XI.) by drawing the oblique circle A B Z. Take the measure of the angle which is to be at the primitive = 45° from the scale of semi tangents, and sweep it round the pole of the primitive (from the centre) as *s t*.

Find the pole, *x*, of the oblique circle (Prob. VII.), and round *x* draw a small circle equal to the other given angle, or 50°, as *e f*, from scale of semi-tangents. Where the two small circles cut, as at *n*, will be the pole of the oblique circle, which shall make an angle of 45° with the primitive and 50° with the other oblique circle.

Through *o n* draw a diameter ; measure *o n* on the scale of semi-tangents, and lay off its complement beyond *o* ; say to *v*.

Through the three points D*v*C (D C being at right angles to G H) draw the oblique circle D*v*C (Prob. IV.), and A B C shall be the triangle.

To **measure** the parts required : —

1. A C is measured on the scale of chords.
2. A B   „   by Prob. XII.
3. B C   „   by Prob. XII.

By calculation : —

RULE.—From half the sum of the three angles take each of the angles next to the side required. Add the co arcs of the sines of the two angles which adjoin the required sides, together with the cosines of the two remainders. Then half the sum of these logarithms will equal half the cosine of the side required.

Find side B C.

$$\angle A = 130°$$
$$\angle B = 50 \quad \text{sine . . co ar } 0\cdot 115746$$
$$\angle C = 45 \quad \text{sine . . co ar } 0\cdot 150515$$

2)225	
112½	
$\angle B$ = 50	
62¼	1st remainder co sine 9·664406
112½	
: C = 45	
67½	2nd remainder co sine 9·582840
	2)19·513507
cosine 34° 50' =	9·756753
2	
69 40	
180	
110 20 = side B C	

Find side A C by rule of sines (78).

as sine ∠A 130°	.	.	co ar 0·115746
to sine BC 110° 20'	.	.	9·972986
so is sine ∠B 50°	.	.	9·884254
to sine side AC	.	.	9·972986 = ∠AC 70°

Find side AB (78).

as sine ∠A 130°	.	.	co ar 0·115746
to sine BC 110° 20'.	.	.	9 972986
so is sine ∠C 45°	.	.	9·849485
to side AB	.	.	.	9·938217 = ∠AB 60° 9'

107. All the rules necessary for calculating a spheric triangle have been explained. Other formulæ might have been added, but where the application of the subject to practice is the main object sufficient has been given. But in order to obviate any possible difficulties which a student might at first encounter in the application of what has been said, we shall now return to the examples which were constructed from supposed observation (page 65), and show the manner of calculating the results; and the more especially is this necessary, because, in spherics, the triangle drawn as a problem in nautical astronomy differs from that which is more adapted to calculation; and again this, if an oblique triangle, requires some management in order to so construct the triangle as to render it convenient for letting fall an available perpendicular to be used with the "5 circular parts," or Napier's rules. An example (No. 2) will be given in illustration of this difference, while the others will be worked without a perpendicular, leaving it to the student to exercise his ingenuity in further construction or calculation.

108. EXAMPLE 1.—To find the latitude of a place :—

Given, meridian altitude 60°   (The observer north of the sun)
  „   declination   . 20° N.

This question needs no other figure than that already

given (fig. 40), because the sun being on the meridian or
primitive, its solution is a mere question of length of an
arc, no triangle is formed. The sun's altitude being known,
together with its declination, fixes the position of the
equator, and as the pole is always 90° distant from it, and
as latitude is the height of the pole above the horizon, latitude will be the complement of the height of the equator
above the horizon, and is measured on the primitive.

Thus, in Fig. 59, if H d be the altitude   = 60°
and E d the declination = 20° N.

height of the equator = 40

90

∴ E Z or P R = the latitude = 50°

Fig. 59.

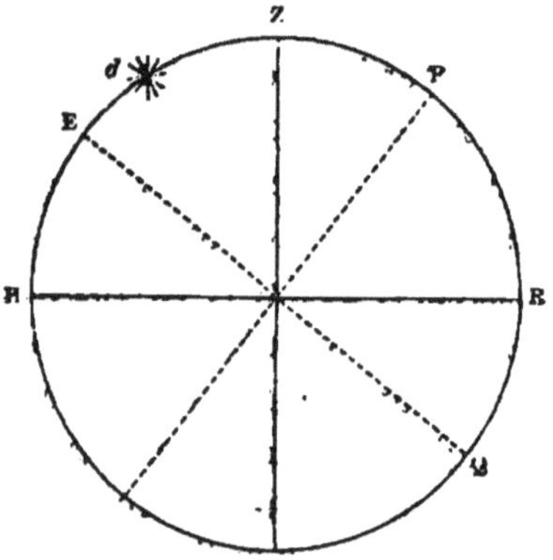

109. EXAMPLE 2.—To find the latitude :—

Given, sun's altitude   50°
   ,,   declination   20° N.
   ,,   azimuth . S. 45° E.

1. By special rule (page 73). It is generally more convenient to use this rule, because it applies to the astro-
nomical figure at one, while to use a perpendicular and
the circular parts requires a reconstruction of the figure,
(which, see onward).

Find ∠ C by rule of sines (79).

as sine side B C 70°	.	.	. co ar	9·027014
to sine ∠ A 135° .	.	.	.	9·849485
so is sine side A B 40°	.	.	.	9·808067
to sine ∠ C .	.	.	.	9·684566 = 28° 56′

side 70	∠ 135°·0′
side 40	∠ 28·56
2)110	2)106·4
½ sum . 55	53·2 = ½ diff ∠ s.

2)30

½ diff . 15

as sine ½ diff of sides 15°	.	. co ar	0·587004
to sine ½ sum 55° .	.	.	.	9·913365
so is tang ½ diff 53° 2′ .	.	.	.	10·123411
to tang ½ contained ∠ .	.	.	.	10·623780 = 13°·23′

2

B = 26 46

Find side A C by rule of sines.

as sine ∠ C 28° 56′.	.	.	. co ar	0·315342
to sine side A B 40°	.	.	.	9·808067
so is sine ∠ B 26° 46′	.	.	.	9·653558
to side A C .	.	.	.	9·776967 = 36° 45′

90

latitude = 53° 15′

Fig. 60.

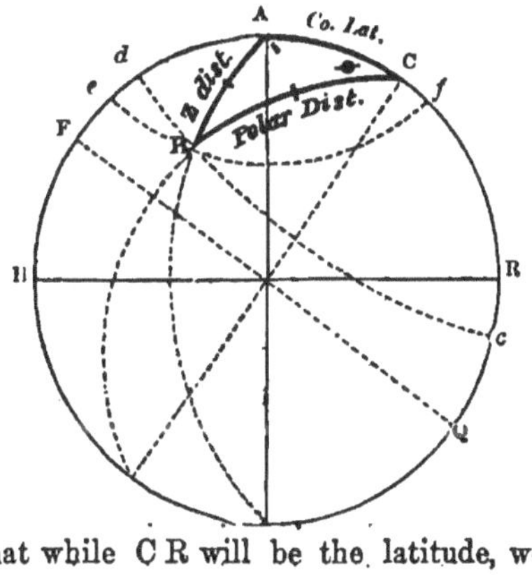

N.B.—Notice that while C R will be the latitude, we, in forming the triangle A B C, use *the complement* of the two given sides, and that the side required in the triangle A B C is therefore the co latitude.

2. *By means of a Perpendicular.*

As previously recommended (90), it is well, as a general rule, to put a given angle at C (by Prob. XI.), and a given side upon A C by scale of chords, letting fall the perpendicular upon B C (or B C produced if necessary) from the angle A by Prob. X.

In this question we have two sides, and their two opposite angles, one of the latter having been found by rule of sines.

The following will be the triangle as constructed for the perpendicular let fall from A :—

             o   *

side A C being 70
side A B ,,   40
∠ C   ,,   28 56
∠ B   ,, 135

Fig. 61.

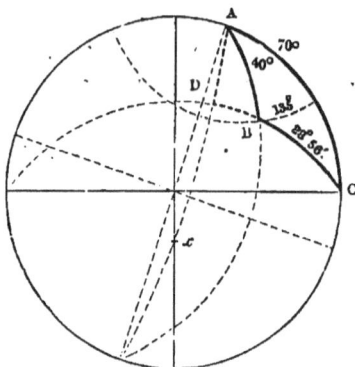

Find D C in △ A D C.

The ∠ C is middle and extremes are conjunct (82).

Then correcting for complements or supplements (83), we have by (87)

as cotang A C 70° . . . . co ar	0·438934
rad	10·
cosine ∠ C 28° 56′	9·942099
to tang side D C	10·381033 = 67° 25′

Find A D in △ A D C.

A·D is middle, and extremes are disjunct, then by (88).

as rad , ...	. 10·
to sine A C 70°	. 9·972986
so is sine ∠ C 28° 56′ . . .	. 9·684658
to sine A D	. 9·657644 = 27° 2′

Find D B in △ A D B.

A B is middle and disjunct, then by (89).

as cos A D 27° 2	. 0·050248 ·
to rad	. 10·
so is cos A B 40°	. 9·884254
to side D B	. 9·934502 = 30° 41

$$\text{side D C} \quad 67 \quad 25$$
$$\text{co lat or side B C} = 36 \quad 44$$
$$90$$
$$\text{latitude required} = 53° \ 16′$$

It would have been shorter (perhaps not so obvious to a beginner) to have found D B by using the △ A D B, and using the complement of A B C (= 45°) as the ∠ D B A, thus:—

The ∠ B 45° is middle and conjunct, then by (87).

as cotang 40°	. 9·923814
to rad	. 10·
so is cos ∠ B 45° . . .	. 9·849485
to tang D B	. 9·773299 = 30° 41′ co lat

110. EXAMPLE 3.—**To find the latitude of a place:**

Given altitude 50° south of the sun
dec   10 S.
time   10 A.M.

In this example, the sun being in south declination and the observer being south of the sun, it is evident that the observer must be in south latitude (remember that latitude is the height of the pole of the sphere above the horizon).

By construction : —

Being in south latitude the south pole will be above
the south part of the horizon. We generally, in construction,
consider the north part of the horizon to lie on the right
of the centre of the figure, and the south part on the
left part.

Assume S the south pole, and draw S O P, and E Q, the
equator at right angles to it.

Draw the great circle S x P (by Prob. XI.), making an
angle of two hours or 30° with primitive.

By Prob. VI., draw the parallel of declination dc at a
distance of 10° south of E Q, or 80° from the primitive.

Fig. 62.

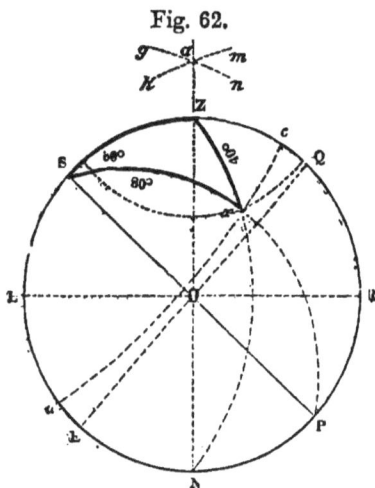

To draw the great circle Z x N, so that Z x = 40° (the co
of 50° the altitude), take the secant of 40 from the centre o,
and sweep an arc as $g\,n$, and the tangent of 40° from the
point x, and sweep another arc $h\,m$ across it ; the intersection will be the zenith line O Z a (Prob. VI.). With the
three points Z x N, describe the great circle Z x N (Prob.
IV.), and S Z x is the triangle, and S Z the co latitude, S H
the latitude required.

By calculation : —

Find $\angle$ Z, by rule of sines (79).

as sine of side Z X = 40° .   . co ar   0·191933
is to sine side of $\angle$ S = 30 .   .   .   9·698970
so is sine side S x — 80 .   .   .   9·993351

to sine $\angle$ Z   =   .   .   .   9·884254 = 50°
    (obtuse by construction)   180°
                              130° = $\angle$ Z

To find $\angle$ x.

a side Z x	=	40° .	. 40°	an angle	130°
a side S x	=	80 .	. 80	an angle	30
		2)120	2)40		2)100
half sum sides =		60	½ diff 20		50 half diff 2 angles

By rule (page 73) :—

as sine ½ diff 2 sides 20° .   .   . co ar   0·465948
is to sine ½ their sum 60° .   .   .   .   9·937531
so is tang ½ diff 2 angles 50° .   .   .   10·076186

to co tang ½ then contained $\angle$.   .   .   10·479665 = 18° 20′
                                     $\angle$ x = 36° 40′

To find co lat S z, by rule of sines (78).

as sine $\angle$ S = 30° .   .   . co ar   0·301030
is to sine side Z x = 40°   .   .   9·808067
so is sine $\angle$ x = 36° 40′   .   .   9·776090

to sine side S Z   .   .   .   .   9·885187 = co lat 50° 9′ S.
                                         90
                               lat 39° 51′

111. EXAMPLE 4. — To find the latitude by a celestial body on the meridian below the pole :

Given, altitude of α Lyræ on meridian below the north pole = 20°
     declination of do.   .   .   .   .   .   . = 38½ N.

This forms no triangle, the star being on the primitive.

Let H R be the horizon. Lay off the alt 20° to a, the star's place. The star being 38½ N. of the equator (in declination), lay off 38½ from a to Q. Draw Q E and P S, and P R will be the latitude, and will equal 90° — (38½° — 20°) = 90° — 18½ = 71½ north.

Fig. 63.

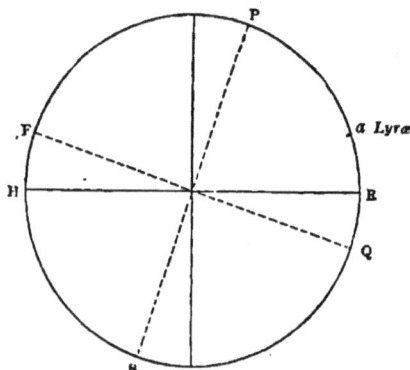

112. EXAMPLE 5. — To find the latitude of a place : —

Given, app. time 9 a.m.
declination, 20° N.
azimuth, S. 60° E.
Observer N. of sun.

Fig. 64.

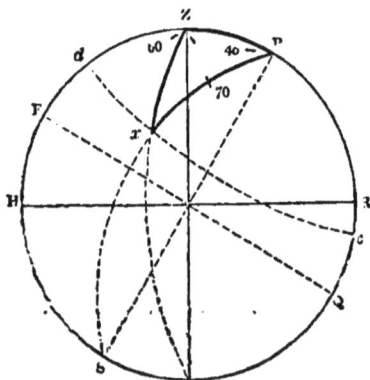

By construction :—

Assume a point P on the primitive, and draw PS and EQ. Draw the oblique circle Px, making 45° with the primitive (Prob. XI.). About P draw the parallel circle dc 70° (the co declination) distant from P (Prob. V.). Through the intersection x draw Zx, making the angle at the primitive = 60, the azimuth from noon, or south (Prob. XI.). Then ZxP will be the spheric triangle, and ZP the co latitude.

By calculation :—

To find Zx by rule of sines (78).

60°	as sine ∠Z 120°	. .	co ar 0·062469
186	is to sine side xP 70°	. .	9·972986
∠x=120	so is sine ∠P 45°	. . .	9·849485
	to sine side Z x	. . .	9·884940 = 50° 6′

To find ∠x (page 73).

sine	50° 6′	∠	120°
side	70	∠	45
	2)120 6		2)75
half sum	60 3		37 30 half diff
	2)19. 54		
half diff	9 57		

as sine ½ diff 2 sides 9° 57′ .   co ar 0·762485
to sine ½ sum 2 sides 30° 3′ . .   9·937749
so is tang ½ diff 2 angles 37° 30′ .   9·884980
to cotang ½ cont ∠x . .   10·585214 = 14 34′
                                                      2
                                        ∠x 29   8.

Find side ZP by rule of sines (78).

as sine ∠Z 120°   ,   .   co ar 0·062469
is to sine side Px 70° .   . .   9·972986
so is sine ∠x 29° 8′   ,   . .   9·687389
to sine of side ZP   ,   ∴   ∵   9·722844 = 31° 53′
                                                      90
                            latitude ZP   58   7 N

113. Example 6.—To find the time (or hour angle).

Given, latitude 45° 40′
declination 10° N.
azimuth S. 45 E.

$$
\begin{array}{lrr}
\text{lat} & 45° & 40′ \\
& 90 & \\
\hline
\text{co lat} & 44 & 20
\end{array}
$$

Fig. 65.

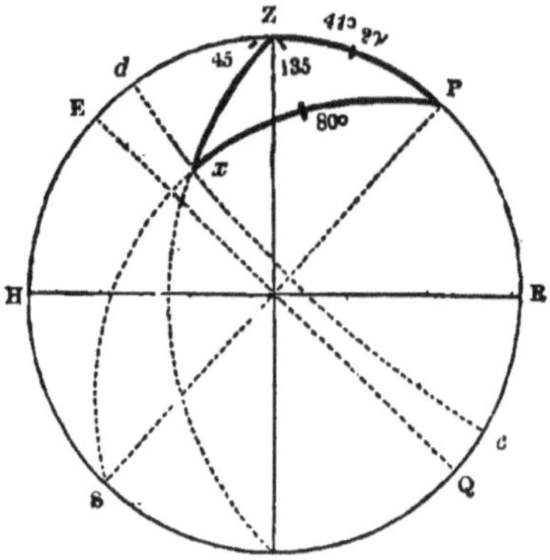

By construction :—

Lay off the lat 45° 40′ from R to P (scale of chords), and
draw diameters at right angles from P (the pole of the
world) draw the parallel d, c with the co declination = 80°
(Prob. V.). Draw Zx, making the angle 45° with the primitive (Prob. XI.); and through the intersection at x draw
(Prob. IV.) xP, and ZxP will be the triangle and ∠P the
hour ∠ required.

By calculation : —

Find ∠ x by rule of sines (79).

$$
\begin{array}{lll}
\text{as sine side } x\text{P } 80° & \text{co ar } & 0\cdot006649 \\
\text{to sine } ∠ \text{Z } 135° & & 9\cdot849485 \\
\text{so is sine side } Z\text{P } 44° \, 20′ & & 9\cdot844372 \\
\text{to sine } ∠ x & & 9\cdot700506 = 30° \; 7′
\end{array}
$$

side . . .	80° 0	∠ 135° 0 0	
side . . .	44 20	∠ 30 7 0	
	2)124 20	2)104 53 0	
half sum . .	62 10	52 26 30 half diff ∠ s	
	2)35 40		
half diff . .	17 50		

Find ∠ P (page 73).

as sine of half diff sides 17° 50' . co ar 0·513925
to sine of half sum 62° 10' . . . 9·946604
so is tang half diff ∠ s 52° 26½' . . 10·114104
to cotang half cont ∠ P . . . . · 10·574633 = 14° 55'
                                                      2
                                hour angle ∠ P = 29   50

N.B.—To convert space into time and the reverse, use the following rules :—

RULE.—Multiply space by 4 and divide the degrees by 60, thus :—

the arc   29° 50'
                    4
        60)119   20
        1h 50m 20s = the above hour angle

To convert time into space :—

RULE.—Reduce hours to minutes and divide by 4, thus :—

the time   1h 59m 20s
                60
           4)119   20
             29° 50'

114. EXAMPLE 7.—To find time.

Given, declination, 20° S. (being in N. lat.)
         altitude   20
         azimuth   S. 45° W.

By construction :—

From Z draw ZxN, making an angle 45° with the primitive (Prob. VIII.). About Z draw the parallel e'f at a

distance of co alt 70° from it (Prob. VI.). Through their
intersection x draw a parallel of declination $dc=$to polar
distance 110 ($=90°+$dec. 20°) (Prob. VI.); and draw an
oblique circle through P x S. (Prob. X.) Then Z x P will
be the triangle and the $\angle$ P the hour angle required.

By calculation :—   180°

Find $\angle$ P by rule of sines. $\angle$ D Z x 45

     as sine side 110°   co ar 0·027014 $\angle$ x Z P $\overline{135°}$
     is to sine $\angle$ 135° .   .   9·849485
     so is sine side 70°   .   9·972986
     to sine $\angle$ P   .   .   . $\underline{9·849485}$ = hour $\angle$ P   45⁾

                                   4 (page 97)
                               $\overline{60)180}$
                                 3 hours

N.B.—3h. p.m., because sun was west of meridian.

Fig. 66.

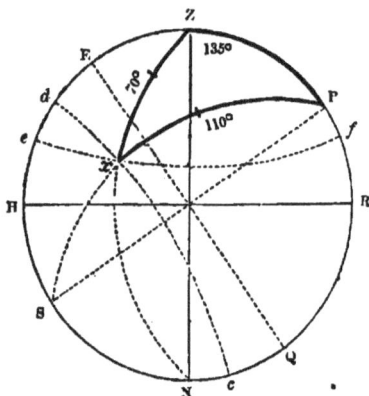

115. EXAMPLE 8.—To find time.

         Given, latitude   21° N.
               declination   20° S.
               altitude   30

By construction :—

Lay off P R from the scale of chords = the latitude 21°

(considering P, as usual, the north pole of the world). About S, the south pole, draw a parallel equal to sun's north polar distance, or 110° (Prob VI.). About Z, the zenith, draw a parallel fg equal to co alt (or zenith distance) (Prob. VI.). Through ZxN draw an oblique circle (Prob. IV.) and ZxP will be the triangle and ∠ P the hour angle.

Fig. 67.

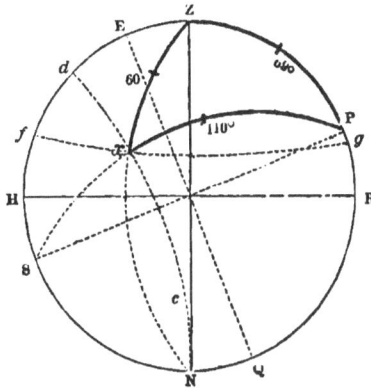

By calculation :—

N.B.—Three sides are given ; find ∠ P (page 84).

```
a side Zx   .  =   60°
a side ZP   .  =   69    co ar sine   .  0·029848
a side xP   .  =  110    co ar sine   .  0·027014
                 2)239
half sum        119 30'
ZP               69
                ─────
                 50 30 = 1st remr   .  9·887406
                119 30
                110
                ─────
                  9 30 = 2nd remr   .,  9·217609
                                   . 2)19·161877
                                    ─────────
       ·sine        22° 24' = 9·580939
                               2
       ∠P            44  48
                              4
                    ─────────
                 60)179  12
  time from noon        2h 59m 12s
```

116. EXAMPLE 9.—To find an azimuth.

Given, latitude   36° S.
declination 20° N.
altitude   20°

By construction:—

Place S, the south pole of the heavens, 36° above H, the assumed south part of hori-

Fig. 68.

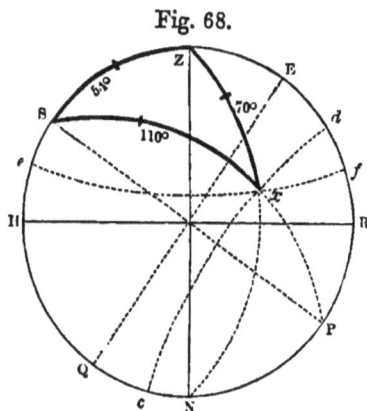

zon Z. Draw diameters as usual. About P, the north pole, draw the parallel $d\,c$ at a distance equal to the co declination 70°(Prob.VI.). About Z draw the parallel $e\,f$ equal to the co alt 70° (Prob. VI.). Through the point x, where these cut, draw ZxN and SxP (by Prob. IV.) and ZxS is the triangle, and angle Z the required azimuth.

By calculation:—

N.B.—Three sides are given. Find $\angle$ Z (page 84).

```
        a side SZ 54°    sine  .    co ar 0·092042
        a side Sx  110
        a side Zx   70    sine  .    co ar 0·027014
                 2)234
                   117
        SZ         54
                   63 1st  remainder sine 9·949881
                   117
        Zx          70
                    47 2nd remainder sine 9·864127
                            2)19·933064
              sine 67° 48'  =   9·966532
                           ‖2
           azimuth ∠ Z  135  36
```

N.B.—S Z and Zx are the two sides which make the required angle, therefore *their* co ar sines are used.

117. EXAMPLE 10.—To find an azimuth.

Given, latitude   21° N.   declination   20 S

time   9 a.m.   90

declination   20° S.   110 = polar distance
                                                      from N

By construction :—

Place the north pole
at P, 21° above the hori-
zon R. Draw diameters.
Draw the oblique circle
PxS with the ∠ 45° from
the primitive(45° = 3 hrs.)
(Prob. VIII.). About S,
the south pole, draw a
parallel equal to the co declination (by Prob. VI.),
and through the intersec-

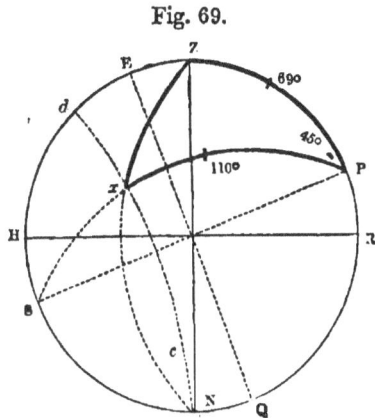

Fig. 69.

tion x draw (Prob. VI.) ZxN and ZxP will be the triangle,
and ∠ Z the azimuth required.

By calculation :—

We have two sides given and an included angle.

a side xP	110		
a side ZP	69	included ∠ 2)45°	
	2)179		
		22 30 half contained ∠	
half sum	89 30		
	2)41		
half sum	20 30		

To find the other angles (page 79).

as sine ½ sum 2 sides 89° 30′	.	co ar	0·000017
to sine ½ diff 2 sides 20 30	.	.	9·544325
so is cot ½ contd ∠ 22 30	.	.	10·382776
to tang ½ diff other angles	.	.	9·927119 = 40° 13′
as cos ½ sum 2 sides 89° 30′	.	co ar	2·059158
to cot ½ diff 2 sides 20° 30′	.	.	9·971588
so is cot ½ contd ∠ 22 30	.	.	10·382776
to tang ½ sum other angles	.	.	12·413522 = 89° 47

sum is the azimuth greater ∠ Z   130 00

less ∠ x 49 34

118. EXAMPLE 11. — To find a ship's course when sailing on a great circle, and the distance between port and port.

Given, ship's latitude in . . . . . . 50° N.
            latitude of place bound to . . . 10° N.
       the difference of longitude between the two places 60° W.

The term great circle "course" is deceiving, inasmuch as no part of a circle is a straight line. A ship could not sail upon a great circle without constantly changing her course by compass; and, therefore, "great circle sailing" is positively impracticable, because a great circle "track" cuts no two meridians at an equal angle. The passage of a ship along a great circle track is evidently a series of courses *tangential* to it.

Fig. 70.

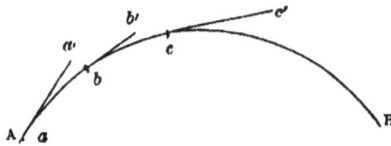

Suppose, in the above figure, A B to be a given great circle track; upon a ship's starting at *a* for the point B a compass course, if long continued, would take her considerably away from her proper track, and, we will say, place her at *a'*. It is evident, therefore, that the *shorter these compass courses are made, the more will the ship keep to her true course,* and the shorter will be the distance required to be sailed over. As an instance: A ship was 12 hours in going from *b* to *b'*, and 24 hours in going (at the same rate) from *c* to *c'*; we find that in these cases she would be leaving her great circle track altogether. It remains, therefore, to provide a *ready method* of finding how to steer by compass so as to depart *as little as possible* from our proper track, which, of course, would be the nearest distance between the port left and port bound to.

The term *tangent* sailing is the only correct designation of this method, and was first suggested to the Astronomer Royal by the author in 1857.

From the above example to find the first or initial tangent course, we proceed thus—

By construction :—

Consider the diagram as a hemisphere drawn on the plane of a ship's meridian (*the ship being somewhere on the meridian*). P S would represent the poles, and E Q the equator. Let L. represent the latitude, or the

Fig. 71.

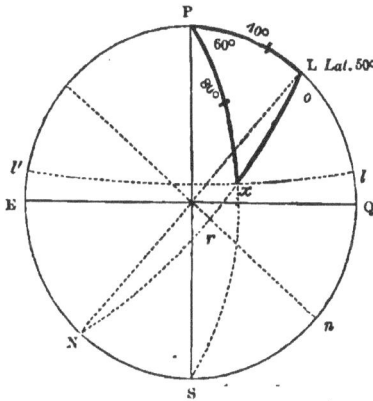

ship's place, and l the latitude of port bound to, or L Q = 50° and lQ = 10°. Draw diameters to point L, and draw PxS, making an angle of 60° (the difference of longitude) with the primitive (Prob. VIII.). Draw the parallel circle $l\,l'$; and through the point of intersection x draw the oblique circle LxN, Prob. X., and the $\angle$ xLQ will be the tangent course required ; measured on $r\,n$ (being 90° distant) it equals 63° 17′.

By calculation :—

We have here two sides and an included angle.

```
a side      80      included ∠ 2)60°
a side      40                 30  = half contained ∠
         2)120
half sum   60
          2)40
half diff  20
```

To find the other angles (page 79.)

```
as sine ½ sum of sides 60° .    .    co ar 0·062469
to sine ½ diff 20°      .    .    .    9·534052
so is co tang ½ contained angles 30°  .  10·238561
to tang ½ diff angles  .    .    .    9·835082 = 34° 22′

cos ½ sum of sides 60° .    .    .    co ar 0·301030
to cosine ½ diff 20°   .    .    .    9·972986
so is cotang ½ contained angle 30°  .  10·238561
to tang ½ sum other angles   .    .    10·512577 = 72° 55′
                                  ∠ L      107   17
                                           180
                  the tangent course ∠ xLQ   72   43
                                  or S 72   43  W
```

119. EXAMPLE 12.—To find the distance between port and port upon a great circle (in the above example).

Find side Lx by rule of sines (78).

```
as sine ∠ L (as above) 107° 17′ .  co ar 0·020066
is to sine side 80°   .    .    .    9·993351
so is sine ∠ P 60°    .    .    .    9·937531
to sine side Lx .    .    .    .    9·950948 = 63° 17′
                                           60
          distance to make good Lx  3797 miles.
```

Suppose that, three days afterwards, the ship was in lat, not 50° but 45° N, and diff of long between the two places was not 60°, but 50°, what would be her altered course?

Working as above would give the course about 67°, and the distance about 55°, or 3300 geographical miles.

N.B.—By Saxby's spherograph (a simple instrument which thoroughly illustrates nautical astronomy, and works any spheric triangle without calculation) a great circle, or tangent course, is easily obtained in five to ten seconds; being thus rendered more simple than even a Mercator's course.

Fig. 72.

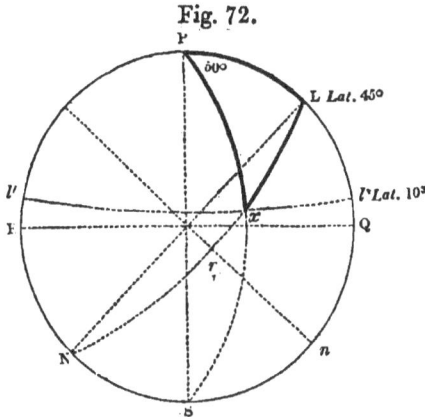

120. Example 13.—To find the "latitude of vertex" of a great circle track.

Given, latitude of the ship   36° S.
   "   bound to   40° S.
   difference of long.   110°

Fig. 73.

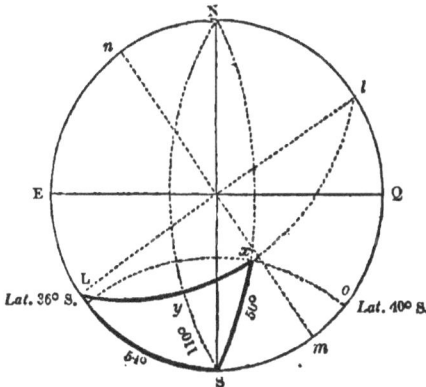

By construction :—

Make E L equal to lat 36° (from the scale of chords)
south of the equator, and L will be the ship's place on her
meridian.   Draw diameters L l and $m n$.   About the south
pole, S, draw a parallel circle L o, distant 50° the co lat
(Prob. V.).   Draw the oblique circle S x N, making 110° with
the primitive (Prob. VIII.).   Through the intersection x draw
L x l, and L x S will be the triangle.   Let fall a perpendicular
S y upon L x (the great circle track) (by Prob. X.) from
the pole S, and y will be the part of the track nearest to
the south pole, or the *vertex*.   Measure S y (by Prob. XII.),
and its complement will be the latitude of vertex.

By calculation :—

To find the two angles.

We have two sides and the included angle given.

```
a side      54°
a side      50      included ∠ 2)110°
         2)104                    55   ½ contained ∠
half sum    52
half diff  2)4
            2
```

To find the other angles (page 79).

```
as sine ½ sum of sides 52°   .      .    co ar 0·103468
to sine ¾ diff of sides 2°       .      .        8·542819
so is cotang ½ contained angles 55°      .       9·845227
to tang ½ diff other angles   .      .      .    8·491514 = 1° 46′ 30″

as cos ½ sum of 2 sides 52° .      .    co ar 0·210658
to cos ½ diff 2 sides 2°        .      .         9·999735
cotang ½ contained angles 55°      .      .      9·845227
to tang ½ sum of other angles      .      .    10·055620 = 48° 40′ 00″

                                        sum = ∠ x   50  26  30
                                        diff = ∠ L   46  53  30
```

Find S y by rule of sines (78).

```
as rad        .       .      .   10·000000
to sine side 50° .      .    9·884254
so is sine ∠ x 50° 26′ 30″   9·887041

to sine side S y .      .    9·771295 = 36° 12′
                                           90
              latitude of vertex " y "    53   48
```

121. EXAMPLE 14.—To find an altitude of a celestial body.

Given latitude 30° N.
   „   time   11 A.M.
   „   declination 20° S.

Fig. 74.

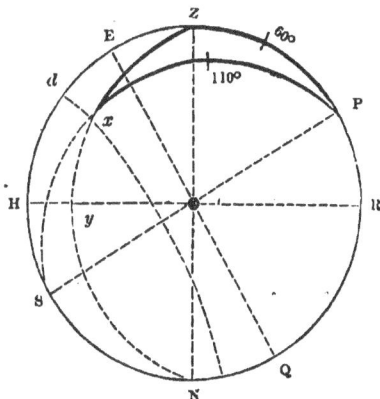

By construction :—

Lay off the lat from R to P by scale of chords.

Draw the hour circle 11 A.M. = 15° by Prob. VIII. = P x S.

Draw the parallel of declination $d\,e$ by Prob. VI.; where these intersect will be x the altitude sought.

Through Z x N draw a great circle (Prob. IV.), and $x\,y$ will be the measure of the altitude (Prob. XII.), and P x Z is the triangle.

By calculation :—

With the co lat Z P = 60°, the pole distance x P = 110, and the hour $\angle$ P = 15°, we have two sides and an included angle.

By rule page 79 :—

```
        a side  110°
        a side   60          2)15°
               2)170 sum      7° 30′ half contained ∠
                 85 ½ sum
               2)50 the diff
                 25 ½ diff
```

To find ∠ Z.

as sine of ½ sum of 2 sides 85°	. co. ar.	0·001656
is to sine ½ diff 2 sides 25° .	. .	9·625948
so is cotang ½ contained ∠ 7° 30′ .	.	10·880571
to tang ½ diff 2 angles . . .	.	10·508175 = 72° 45′ 30″
as cos ½ sum 2 sides 85° . .	. .	1·059704
is to cos ½ diff 2 sides 25° .	. .	9·257276
so is cotang ½ contained ∠ 7° 30′ .	.	10·880571
to tang ½ diff 2 angles . . .	.	11·897551 = 89° 16′ 30″

∠ Z   162   2

To find Z x the zenith distance (78).

as sine ∠ Z 162° 2′ . . .	. co. ar.	0·510796
is to sine side 110° . .	. .	9·972986
so is sine ∠ P 15° . .	. .	9·412996
to sine side Z x . .	. .	9·896778 = 52° 30′

90

altitude = x y   37° 30′

122. Example 15.—To find an amplitude.

Given latitude   50° N.
   „   declination 20° N.

Fig. 75.

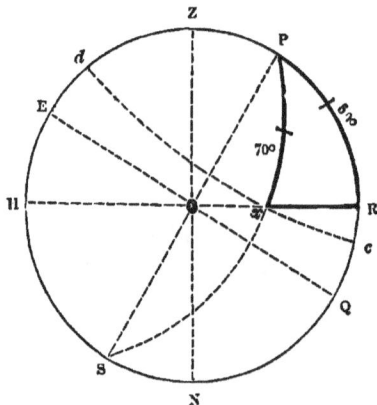

By construction:—

Lay off the lat R P = 50° (scale of chords).
Draw P S and E Q.

Draw a parallel of declination $dc = 20$ N. from E Q (Prob. VI.).

Through point of intersection x draw the great circle P x S (Prob. IV.), and P x R will be the triangle, and x o the amplitude.

By calculation :—

P x will be a middle part (83), and the extremes are disjunct.

To find x R (89).

as cos 50°	.	.	.	.	co ar.	0·191933
is to rad	.	.	.	.	.	10·
so is cosine 70°	.	.	.	.		9·534052
to cosine x R	.	.	.	.		9·725985 = 57° 51′ azim

$$\frac{\qquad}{90}$$

the amplitude x o = 32° 9′ N.

N.B.—The ordinary rule is, add the secant of the latitude to the sine of the declination, and the sine of the sum is the amplitude.

Thus,—sec 50° . . . 0·191933
            sine 20 . . . 9·534052
                        ‾‾‾‾‾‾‾‾‾‾
                        9·725985 = 32° 9′

Remember the secant is the reciprocal of the cosine, and the cosine of 70° = the sine of 20°; hence the rule.

123. EXAMPLE 16.—To find the time of daybreak.

N.B.—Daybreak is the time at which the sun's centre is just 18° below the horizon of the place.

Given latitude . . 50° N.
    „ sun's declination . 9° 43′ S.

By construction :—

Lay off the lat 50° N. from R to P from scale of chords. Draw the parallel of declination d c by Prob. VI.

Draw the parallel tw of 18° below H R the horizon by Prob. VI.; where the parallels cut at x is the sun's place at daybreak.

Fig. 76.

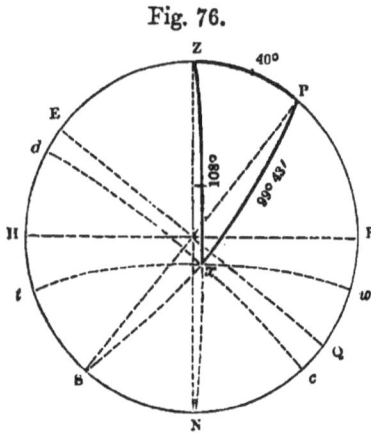

Draw the great circle P x S (the hour circle), and ZxN the azimuth by Prob. IV., and Z x P will be the triangle, and the $\angle$ xP R the time from midnight.

By calculation :—

Here are three sides given, viz., the co lat 40°, the polar distance 99° 43′ and the side Z x = 90° + 18° = 108°.

By rule, page 84.   To find $\angle$ P.

```
a side 108°              co arc sine  .    .    . 0·006275
      99 43′             co arc sine  .    .    . 0·191933
      40
   2)247 43
     123 52 ½ sum   .
      99 43
      24  9             1st remainder sine  .  9·611858
     123 52
      40
      83 52             2nd remainder sine  .  9·997507
                                              2)19·807573
           ½ ∠ Z P x =  53° 15′   =    9·903786
                           2
           ∠ Z P x =  106  30
                       180
           ∠ R P x =   73  30   in space
                        4   by rule page
                   60)294  00
apparent time of daybreak    =     4ʰ 54ᵐ in time
```

N.B.—To be corrected by the equation of time for the day as given in almanacks, so as to get mean time.

124. EXAMPLE 17.—To find the time of rising of a celestial body.

Given latitude of place . . . 51° 27′ N.

„ declination of the object . 10 15 S.

Fig. 77.

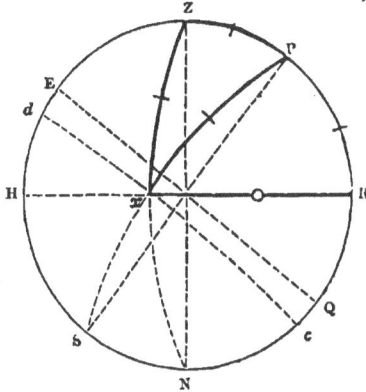

By construction :—

Lay off the latitude at R P.

Draw the parallel of declination dc by Prob. IX.

Where this cuts the horizon will be the place of the sun's rising, as at x.

Draw P x S, Prob. IV., and x P R will be the triangle, and the ∠ x P R will be the time of rising. P R Q, &c., being the midnight meridian.

By calculation :—

To find ∠ P.

∠ P is middle and extremes are conjunct (83).

Then by (86) :—

As rad .	.	.	.	.	.	. 10·
is to cotang P x 100° 15′ .		.	.	.	. 9·257269
so is tang 51° 27′ .		.	.	.	. 10·098617
to cosine ∠ P	.	.	.	.	. 9·355886 ▪ 76° 53′

$$ 4$$

$$ 60)307\ 42$$

time of sunrise . ▪ 5ʰ 7ᵐ 42 S.

If we use the triangle Z *x* P, we have a quadrantal triangle for Z *x* = 90°.

By rule (page 69).

Find ∠ Z P *x*.

as rad . . . . 10·
is to cotang Z P 38° 33′ . . . . 10·098617
so is cotang P *x* 100° 15′ . . . . 9·257269
to cosine ∠ Z P *x* . . . . . 9·355886 = 76° 53′ &c.

The preceding examples comprise all ordinary questions to which the attention of the navigator is likely to be called. When a student has read this little book he will be better able to comprehend works written upon the subject of spherics, in which the *theory* is explained. It is more than probable that a small volume may shortly follow the publication of this, adapted to those who, not content with an "*initiation*" into nautical astronomy, desire to become further acquainted with so beautiful, enticing, and useful a branch of study.

THE END.